## DATE DUE

| | |
|---|---|
| FEB 2 8 2000 | |
| NOV 0 2 2004 | |
| AUG 2 0 2010 | |
| | |
| | |
| | |
| | |
| | |
| | |
| | |
| | |
| | |
| | |
| | |
| | |
| | |

BRODART, CO.                    Cat. No. 23-221-003

# BASIC TRAINING: WHAT TO EXPECT AND HOW TO PREPARE

# BASIC TRAINING: WHAT TO EXPECT AND HOW TO PREPARE

*by*

Robert F. Collins, Col. (ret.)

THE ROSEN PUBLISHING GROUP, INC.
New York

Published in 1988 by The Rosen Publishing Group, Inc.
29 East 21st Street, New York, NY 10010

*First Edition*

**Library of Congress Cataloging-in-Publication Data**

Collins, Robert F., 1938–
    Basic training: what to expect and how to prepare/by Robert F. Collins.
        p.    cm.
    Includes index.
    ISBN 0-8239-0833-X
        1. Military education—United States—Basic training—Juvenile
    literature.    2. United States—Armed Forces—Vocational guidance—
    Juvenile literature.    I. Title.
    U403.3C65    1988
    355.5′4)0973—dc19                                                   88-25234
                                                                        CIP

*Manufactured in U.S.A.*

This book is dedicated to the millions of young men and women who have successfully completed basic training and gone on to serve their country with honor, dignity, and courage.

# About the Author

Robert F. Collins is a recently retired officer who served more than twenty-five years in the US Army. He enlisted in the Army in 1960, graduated from Officer's Candidate School in 1962, and achieved the rank of Colonel before his retirement in 1985. Col. Collins was in the Military Intelligence branch and served tours of duty in Korea, Vietnam, Germany, and the United States. He is a Soviet Foreign Area Officer with extensive travel in Eastern Europe and the Soviet Union. He taught US National Security Policy at the US Army's Command and General Staff College for five years and was a Professor of Military Science for two years. His decorations include the Army Commendation Medal, the Meritorious Service Medal, the Bronze Star, and the Legion of Merit.

# Acknowledgments

The author wishes to acknowledge the assistance provided by the representatives of the Army, Navy, Air Force, Marine Corps, and Coast Guard. These representatives were extremely helpful and cooperative in providing their opinions and furnishing military publications for use as source material. Much of the material in this book has been drawn directly from military publications to insure both accuracy and currency of the statements. Errors in the book, of course, are mine.

# Contents

# Introduction

The Department of Defense requires an input of over 300,000 voluntary enlistees each year to fulfill service requirements. Before 1973 the United States government did not have to rely exclusively on enlistments to meet its manpower goals; rather, the vast majority of new members entering the armed forces were drafted into the Army or joined other services because they were subject to the draft. However, the draft was abolished in 1973.

Consequently, since that time each service has had to design and implement programs to attract the number and type of people that it requires to accomplish its mission. As a result, a variety of programs and opportunities are available to a young man or woman who is considering enlisting in the Army, the Navy, the Air Force, the Marines, or the Coast Guard. Although this book is designed primarily to introduce you to and prepare you for basic training in each of the services, another purpose is to emphasize the importance of obtaining as much information as possible about each service before making any commitment to an individual service. You are well advised to talk to as many people as possible about the armed forces, including recruiters, teachers, parents, friends, former service members, and active duty people and to consider carefully all your obligations upon entering the service before making a decision to enlist. Obtaining information about each service is not difficult. All the services are eager to inform every prospective enlistee about the opportunities, benefits, obligations incurred, and promotion chances associated with its individual service.

The purpose of this book is to present in one volume information on the basic training programs (sometimes called recruit training) in all the services to assist the young man or woman considering entering the armed forces after high school or while in college or university. It is possible to enlist in the armed forces without a high school education, but all the services encourage applicants to obtain a high school education, and some career fields are closed to applicants without it. All the services are trying to identify, enlist, and properly train qualified young men and women. The services want you to complete the basic training and become an effective, producing member of your service.

No attempt will be made to minimize the difficulties of trading a civilian life for a military life, nor to soften the rigors of discipline required to complete the training to graduation. The young men and women who can adapt to the demands of the military will be physically fit, morally sound, and socially adept. Most important, they will be strongly motivated to serve their country. Potential applicants are advised to examine closely their reasons for wanting to enter the service. Military service is demanding, and attrition rates are significant. Still, the military is an excellent place for a lot of young people to grow and mature. It must be recognized, however, that the military is not for everyone. Some people, for whatever reasons, will never be comfortable nor find self-fulfillment in the military. The military also recognizes this fact and makes every effort to identify people not suitable for service. It is in the best interests of both the individual and the services to identify and separate those individuals as soon as possible.

Many erroneous ideas are held about basic training as well as the military itself. Unfortunately, many are based on what has been presented in post-Vietnam movies and not on reality. The image of brutal drill sergeants in basic training is more a creation of Hollywood than a depiction of real life. The vast majority of drill sergeants in the services are highly motivated, dedicated to their job and country, well trained for their job, and extremely interested in their trainees as individuals. The days of a military peopled by unthinking automatons that blindly follow orders are long gone, if in fact they ever existed. Today it seems that the appeal of military life is on the rise. A renewed spirit of patriotism is evident among American citizens after having reached a low point in the late 1960s and early 1970s during the Vietnam War. Obviously, many other factors contribute to the current sense of pride and desire to serve the country. The end result is that the services seem to be attracting better educated, better qualified, and better motivated applicants than in the recent past. Because the services can be more selective, it is all the more important that applicants be fully knowledgeable about application procedures, physical and intellectual qualifications, and what is expected of them during their initial phases of training.

Enlisting in the armed forces does not mean a twenty-year commitment. Indeed, the overwhelming majority of enlistees serve only the length of their initial commitment, and then return to civilian life having profited greatly from their experiences. Generally speaking, the services require an eight-year commitment, with active duty ranging from two to six years and the remainder of the commitment spent in the Reserve forces. Of course, there are always exceptions, and that is why it is crucial that you understand all contractual obligations before signing any agreement with the government. This does not imply that the services would attempt to have you sign something without your understanding it completely; rather, the opposite is true. All the services scrupulously avoid deceit or even the appearance of deceit in today's enlistment procedures.

It should be noted that enlisting in all the services, plus initial training, benefits, and obligations, is similar in many ways, but at the same time each service has its own special procedures. The first three chapters cover general information relating to all the services, and subsequent chapters cover service-specific information on basic training.

*Chapter* I

# General Military Information

Today's military, besides being the largest employer in the nation, is also the employer of the youngest workforce in the nation. The average age of the enlisted force is the early twenties. This situation results from the tremendous turnover of people in the military each year, and the fact that the military requires young, physically fit personnel to carry out the hazardous and physically demanding tasks associated with it in both peace and wartime. The infusion of new people each year guarantees that the military will not become stagnant, because it must constantly deal with new ideas and new perspectives. Change and adapting to change are essential parts of military service, whether you intend to be a two-year enlistee or a career soldier, sailor, airman, or marine. As a practical matter, one of the most important things you can do to ensure your successful completion of basic training is to be in top physical shape. As is detailed in subsequent chapters, basic training is physically very demanding. If you report in poor shape, you will find yourself at a great disadvantage in trying to complete the training. On the other hand, if you report in great physical shape, you will find it much easier to adapt to the new demands put upon you. It is well worth the time and effort to participate in a rigorous conditioning program before reporting to your basic training station.

Whether you intend to serve only a short time in the military or to make it a career, it is important to understand the place of the military in American society. The military plays a unique role in the American form of government. It is a role that has been shaped by experience, tradition, and the unique American values of individual worth and personal freedom. The tradition of soldiery has not developed in the United States as it has in other countries. Indeed, from the very beginning of the American national experience it has been generally agreed that a large standing army was neither desired nor required. This belief was based on many circumstances. Starting with the colonial experience, the early American settlers, with hope for the future and tremendous optimism, were strongly independent and convinced that they were able to protect themselves. One of the reasons they had left Europe was to escape militarized, regimented, authoritarian societies; the American experience encouraged self-reliance, with security concerns best handled by oneself and one's neighbors. It is only since World War II that Americans have come to realize that there is a valid requirement to have relatively large professional armed forces not only in place but also prepared to fight on short notice if necessary.

The United States has been blessed by its location from a security point of view. Until the beginning of the twentieth century it was relatively invulnerable to attack by foreign powers. Flanked by oceans on both east and west and having friendly neighbors on the north and south, the United States developed its traditions and way of looking at the rest of the world in a secure, relatively isolated manner. An abundance of natural resources, temperate climate, and productive agricultural lands further promoted its independence and self-reliance. The United States did not play a world role until this century, and no large standing military force was required to protect it from invaders, keep the sea-lanes open, or guarantee the freedom of the air. Conflicts were local and usually of short duration. When faced with an emergency, the American people rallied to the call for arms, fought bravely, and attempted to resolve the conflict quickly. Military leaders have emerged in time of crisis and received honor and adulation for their deeds, but the American people have always insisted that the citizen army be disbanded as soon as the crisis was over.

Americans have always been sensitive to the dangers of too strong a military influence on government. One of the bedrock tenets of our democracy, guaranteed by the Constitution, is the civilian control of the military. It is an inviolable rule that the military only carries out policy; the military cannot make government policy. The role of the military is to advise civilian decision-makers and then implement their decisions. A small professional force would be able to carry out most missions requiring military force, and American fighting troops could be mobilized for large-scale conflicts if necessary. This idea worked well for the United States until this century, but now circumstances and the world situation have altered drastically.

The United States today is the acknowledged leader of the free world, with global responsibilities and global obligations. The United States does not covet territory, and it is not militaristic, but it must have an adequate standing military force to protect its own interests and the interests of other democratic nations. Our boundaries, so secure for hundreds of years, are now vulnerable to attack from both the air and the sea. Our frontiers now extend to the Far East, to Africa, Europe, Asia, the Caribbean, and the Indian subcontinent. The world has grown much smaller, thanks to technology and human progress. Events that occur in distant countries now have repercussions that directly affect the United States economically, militarily, and politically. It is no longer accurate to view the world as divided into communist and capitalist camps; other actors play important regional roles, and the balance of power is constantly shifting. Military personnel today not only must be technically and tactically proficient in their fighting skills; they must also have an awareness of and be sensitive to international relations.

During most of the history of this country, the military has been relatively isolated from the civilian community. Military personnel usually were stationed in isolated posts, served in foreign countries, or were stationed in ships at sea. There was no draft until this century, and not all sections of the public were fully represented in the military. As a result, the military knew very little about public concerns and perspectives, and the public in turn knew very little about military customs and traditions. The situation is quite different today. Close and continuing contacts exist between military people and civilians at innumerable levels. The media—newspapers, TV, radio, and movies—provide extensive coverage of military operations, research and development, budgets, training, educational requirements and so on. In our democratic form of government, the people have a constitutional right to be informed of governmental activities—government in the sunshine. The Iran/Contra hearings revealed a great deal of wrongdoing, but the fact that seemed to draw the most ire from the American public was that so many things were done in secret without the knowledge of Congress and the public. Similarly, military undertakings, to be successful, must have the support of the people. Today's military member must understand that America's armed forces are truly a people's armed forces, and that he serves the American people as a whole and not any special-interest group. We learned many lessons from the Vietnam War, but the most important was that no war can be successfully prosecuted without the consent and active support of the US public.

The enlistee today and tomorrow faces unique challenges: how to operate in an environment in which technology advances almost daily and military adversaries have the potential to destroy life. These challenges are all the more difficult because basically it is against the American character to prepare for a future war. We believe that conflicts can and should be resolved by discussion and reason, with force used only as a last resort. We believe in the basic good of humanity and generally that our form of government and way of life are the best. We believe that government must serve the people and not vice versa. These are noble ideas, but we must remember that they are not shared by all governments around the world. Our world view works against programs to keep the military fully prepared on an immediate basis to respond to overt attack. The American tradition and ethic demand that the US can never be an aggressor nation, can never attack first, must conduct warfare in an honorable manner, and must be threatened significantly before resorting to violence. The US public must be kept informed of how the war is being conducted and must be able to see a successful short conclusion to the conflict. These imperatives make service in the armed forces challenging and difficult. However, the main point about these conditions in that they are a genuine expression of the American character, and the military must operate within the bounds of the public's authority and approval.

Today's military members must be prepared to operate effectively anywhere along the spectrum of conflict. This could range from limited unconventional war or combatting terrorism to general nuclear war. The US policy in the nuclear age is one of deterrence. US armed forces must be strong enough to make all potential aggressors realize that the benefits gained from aggression against the United States or its allies will not be worth the risks involved. Potential adversaries must also be convinced that the United States will protect its interests. The likelihood of all-out nuclear war is very low in the foreseeable future; leaders of both the US and the Soviet Union realize that nuclear war has the potential to destroy life on earth. Still, the possibility exists of an unauthorized or accidental firing of a nuclear weapon or even a madman's choice to use a nuclear weapon. Having to cope with such a large range of possibilities imposes difficult burdens on the armed forces.

The highest probability of conflict in the near future is low-level insurgencies and terrorist actions in areas far distant from the United States. An excellent example of the type of operation the US is likely to participate in is its commitment to protect US flag ships involved in the oil business in the Persian Gulf. Although a relatively low-level operation, it still poses dangers to the personnel involved, and the military personnel must be well trained and tactically prepared for the situation. Today's military must be prepared for the possibility of military operations in Europe, Africa, Latin America, the Far East, and the Persian Gulf area. The young man or woman entering the military today must recognize that he or she is setting out on a difficult path. However, most of these people will feel amply rewarded knowing that they are serving their country and helping preserve the American way of life for future generations. Some of the more tangible benefits of military service are briefly touched upon in Chapter III.

*Chapter* II

# Enlistment Information

Making the decision to enlist in one of the military services is an extremely important move and should not be taken lightly. It is a decision that you and you alone should make. You should not make the decision because your parents think it is best for you, or because you can find nothing else you want to do with your life, or because it seems like a good idea on the spur of the moment. If you enlist for these reasons or similar ones your chances for success in the military are greatly diminished. If, on the other hand, you enlist after careful and critical examination of all the information available to you about each of the services, you are off to an excellent start on service in the military that will be both satisfying and professionally rewarding. Developing a satisfying career (whether for two years or thirty years) requires careful planning and informed decision-making. The time you spend gathering information, understanding the consequences of various alternatives, and thinking through your personal preferences will be of enormous benefit as you make career decisions. Over time, you may change your career plans—most people do. But each time you reexamine your career plans you will have more information upon which to base your decisions.

Making a decision is one thing. Carrying out a decision is something else. Most people need help with career decision-making and planning. Your school counselor and parents can help you find information, sort through your options, and make informed decisions. They can also help you make plans to carry out your decisions. However, it is important enough to repeat—the decision to enter the military has to be yours. When beginning to explore your career options, you must consider the facts about all the alternatives open to you. This is particularly true when considering the military as an alternative to civilian employment. Despite many similarities, military and civilian employment differ in important ways.

All people enter the military as either an enlistee or an officer. (Enlistees are often called recruits.) Today's military is the largest employer of high school graduates entering the workforce full time. Each year over 300,000 young men and women, most of whom are recent high school graduates, join the enlisted forces of the Army, Navy, Air Force, Marine Corps, and Coast Guard. The Coast Guard is a bit different from the other services. The Coast Guard is administered by the Department of Transportation during peacetime. During war or other national emergencies, the Coast Guard may become part of the Department of Defense as a component of the Department of the Navy. It is one of the five armed forces, and the same rules and regulations apply in most instances.

Besides being the largest employer in the nation, employing 1.8 million enlisted men and women, the military offers the widest choices of career opportunities. Together, the five services offer training and employment in over 2,000 enlisted job specialties. These 2,000 specialties are divided into twelve broad occupational groups:

- Human services occupations
- Media and public affairs occupations
- Health care occupations
- Engineering, science, and technical occupations
- Administrative occupations
- Service occupations
- Vehicle and machinery mechanic occupations
- Electronic and electrical equipment repair occupations
- Construction occupations
- Machine operator and precision work occupations
- Transportation and material handling occupations
- Combat specialty occupations

Figure 1 shows the distribution of enlisted workers across the twelve occupational groups.

The population of a military base or a naval fleet often equals that of a small to mid-sized city. Like cities, the military needs many services, supplies, and utilities (such as electricity and communications) in order to be self-sufficient. Therefore, the military services have a wide spectrum of occupations. Over three fourths of all military occupations have counterparts in the civilian world

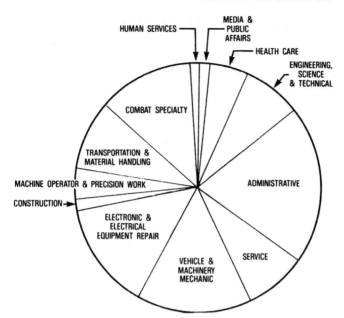

*Figure 1   Distribution of Enlisted Personnel by Occupational Cluster.*

of work. For example, dental hygienist, air traffic controller, computer programmer, aircraft mechanic, and electronic technician occupations exist in both the military and civilian workforces. In fact, this great variety of jobs and the job training provided by the military is the main reason most people give for enlisting. It is certainly true that the military offers some of the finest training available in the United States and has provided a good start for millions and millions of young people. One other point of interest about Figure 1 is that it clearly shows how many support people are required to sustain the combat force. This percentage of combat personnel to support personnel is often called the "tooth to tail" ratio. It points up the fact that the great majority of personnel in the services are in support positions rather than combat positions.

Since 1973 military service has been totally voluntary. Each year the services rely on the enlistment of over 300,000 young men and women to fill military occupational needs. The general qualifications for military enlistment are listed in Table 1. The specific requirements may vary depending on the individual service. Again, it is emphasized that you should talk to many people, especially

**Table 1   General Enlistment Qualifications**

| | |
|---|---|
| **Age** | Must be between 17 and 35 years. Consent of parent or legal guardian required if 17. |
| **Citizenship Status** | Must be either (1) U.S. citizen; or (2) immigrant alien legally admitted to the U.S. for permanent residence and possessing an immigration and naturalization form. |
| **Physical Condition** | Must meet minimum physical standards listed below to enlist. Some military occupations have additional physical standards.<br>Height—For males.   Maximum – 6′8″,<br>                  Minimum – 5′0′′,<br>      For females:  Maximum – 6′8″,<br>                  Minimum – 4′10″<br>Weight—There are minimum and maximum weights, according to age and height, for males and females.<br>Vision—There are minimum vision standards.<br>Overall Health—Must be in good health and pass a medical exam. Certain diseases or conditions may exclude persons from enlistment, such as diabetes, severe allergies, epilepsy, alcoholism, and drug addiction. |
| **Education** | High school graduation is desired by all services and is a requirement under most enlistment options. |
| **Aptitude** | Must achieve the minimum entry score on the ASVAB (Armed Services Vocational Aptitude Battery). Minimum entry scores vary by service and occupation. |
| **Moral Character** | Must meet standards designed to screen out persons likely to become disciplinary problems. Standards cover court convictions, juvenile delinquency, arrests, and drug use. |
| **Marital Status and Dependents** | May be either single or married; however, single persons with one or more minor dependents are not eligible for enlistment into military service. |
| **Waivers** | On a case-by-case basis, exceptions (waivers) are granted by individual services for some of the above qualification requirements. |

your high school counselor and service recruiter, to obtain as much information as possible about your options, opportunities, and obligations. Joining the military involves entering into a legal agreement called an enlistment contract. The service agrees to provide a job, pay, benefits, and occupational training. In return, the enlistee agrees to serve for a certain period of time, which is called the service obligation. The standard service obligation is eight years, which is divided between full-time military duty, called active duty, and reserve duty. Depending on the enlistment program selected, enlistees spend between two and six years on active duty and the balance of the eight-year enlistment period in the Reserves. Enlistment programs vary by service. Major enlistment options include cash bonuses for enlisting in certain occupations, guaranteed choice of job training and assignments, money guaranteed for college after service in the military, and the Delayed Entry Program (DEP). Currently, all services offer a Delayed Entry Program, under which an applicant delays entry into active duty for up to one year. High school students often enlist under the DEP during their senior year and enter a service after graduation. Other qualified applicants choose the Delayed Entry Program because the job training they desire is not currently available but will be within the next year. The enlistment contract specifies the enlistment program selected by the applicant. It contains the enlistment date, term of enlistment, and other options such as a training program guarantee or a cash bonus. If, for whatever reason, the service cannot meet its part of the agreement (for example, to provide a specific type of job training), the applicant is no longer bound by the contract. If the applicant accepts another enlistment program, a new contract is written. The military encourages young people to stay in school and graduate. High school graduates are more likely to be successful in the military than nongraduates. Therefore, the services are now accepting very few nongraduates, and those few who are accepted are strongly encouraged to obtain their high school equivalency while in the service.

Entering the military is a four-step process:

*Step 1: Talking with a Recruiter.* If you are interested in applying for one of the military services, you must talk with a recruiter from that service. Recruiters can provide detailed information about employment and training opportunities as well as answer specific questions about service life, enlistment options, and other topics. They can also provide details about their service's enlistment qualification requirements. If you decide to apply for entry and the recruiter identifies no problems (such as a severe health problem), the recruiter examines your diploma or other educational credentials. The recruiter then schedules you for enlistment processing.

*Step 2: Qualifying for Enlistment.* Enlistment processing occurs at 68 Military Entrance Processing Stations (MEPS) located across the country. At the MEPS, applicants typically take the Armed Services Vocational Aptitude Battery (ASVAB) and receive medical examinations. The ASVAB results are used to determine if an applicant qualifies for entry into a service and has the specific aptitude level required to enter job specialty training programs. If you have taken the ASVAB in high school or post-secondary school, you can use your scores to determine if you qualify for entry into the military services, provided the scores are not more than two years old. Applicants with current ASVAB scores are not required to take the exam a second time.

*Step 3: Meeting with a Service Classifier.* A service classifier is a military career information specialist who helps applicants select a military occupational field. For example, if you were applying for the service, the classifier would inform you of service job training openings that match your aptitudes and interests. Specifically, the classifier would enter your ASVAB scores into a computerized reservation service. Based on your scores, the system would show the career fields and training programs for which you qualify and when job training would be available. After discussing job training options with the classifier, you would select an occupation and schedule an enlistment date. Enlistment dates may be scheduled for up to one year in the future to coincide with job training openings. Following selection of a military training program, you would sign an enlistment contract and take an oath of enlistment. If you had decided to enter the service under the Delayed Entry Program, you would come to an agreement with the classifier about the exact date you would enter the service.

*Step 4: Enlisting in the Service.* After completing enlistment processing, applicants who select the immediate enlistment option receive their travel papers and proceed to a military base for basic training. Applicants who select the Delayed Entry Program option return to the MEPS on their scheduled enlistment date. At that time they officially become enlistees and proceed to a military base. In the uncommon event that your guaranteed training program, through no fault of your own, is not available on the reserved date, you have three options:

- Make another reservation for the same training and return at a later date to enter the service.
- Select another occupation and reserve training.
- Decide not to join the service and be free of any obligation.

The Military Oath of Enlistment is the same for all services:

I do solemnly swear that I will support and defend the Constitution of the United States against all enemies, foreign and domestic; that I will bear true faith and allegiance to the same; and that I will obey the orders of the President of the United States and the orders of the officers appointed over me, according to regulations and the Uniform Code of Military Justice. So help me God.

The military generally provides three kinds of training to its personnel: recruit training, job training, and continuing education. The focus of this book is on basic training, and subsequent chapters detail basic training requirements in each of the services. General information about basic training in all the services follows.

Recruit training, popularly called basic training, is a rigorous orientation to the military. Depending on the service, recruit training lasts from six to eleven weeks and provides a transition from civilian to military life. The services train recruits at selected military bases across the country. Where an enlistee trains depends on the service and the job training to be received. Through basic training recruits gain the pride, knowledge, discipline, and physical conditioning necessary to serve as members of the armed forces. Upon reporting for basic training, recruits are divided into training groups of forty to eighty people, meet their drill instructor, receive uniforms and equipment, and move into assigned quarters.

During basic training recruits receive instruction in health, first aid, and military skills. They also improve their fitness and stamina by participating in rigorous daily exercises and conditioning. To measure their conditioning progress, recruits are tested on sit-ups, push-ups, running, and body weight. All services require members to be physically fit as well as meeting weight standards. In fact, some of the services do not allow enlistees to smoke in basic training as well as discouraging smoking by all members. Recruits follow a demanding schedule. Every day is carefully structured with times for classes, meals, physical conditioning, and field instruction. Some free time (including time to attend religious services) is available during basic training. After completing basic training, recruits normally proceed to job training.

Basic training is physically, emotionally, and mentally demanding, but if you approach the training with the right attitudes you can successfully complete it even if you have had no prior military experience. You must remind yourself that millions and millions of young people have successfully completed the training. You must focus on both short- and long-range goals. Your main goal is to complete basic training, and your short-term goal should be to do the best you can each day. The service wants you to complete the training, and the military cadre will assist you in any way they can. In fact, if you feel that you need extra instruction in any phase of the training, the drill instructors will provide it.

The single most important thing in times of stress—and basic training is certainly a time of stress—is your attitude. In ninety-nine cases out of a hundred, if you think you can do something you will be able to do it. Tell yourself that no matter how tough the training is, you are going to see it through. As soon as you start feeling sorry for yourself or start telling yourself that the training is too difficult, you are headed for trouble. Focus on the positive aspects of your training and how much you are learning. It is only natural that at times you will be discouraged, tired, and frustrated, but just recognize those feelings as passing phases and do not let them dominate your thinking. Focus on the personal rewards and satisfaction you will feel upon successful completion of the training.

You will find you make some lasting friendships during basic training. The military wants you to develop a feeling of teamwork, pride, and camaraderie with the other members of your unit in your military career. You will find that basic training is a lot easier if you develop such friendships. If you can maintain your sense of humor and be enthusiastic about your situation, you are almost guaranteed to be successful. Basic training, like almost everything else in life, will be easier if you work with other people.

It is not my intention to minimize the rigors of basic training, but at the same time, do not enter upon it with the idea that it will be too difficult for your abilities. The military members who provide your training want you to succeed, just as do the members of your platoon, squadron, and company. You must tell yourself that military life is different from civilian life and be willing to learn, adapt, and change. Become interested in what is going on in your training; all services want their members to feel that they are part of the team. No service wants unthinking, unfeeling, or uncaring enlistees. If you have questions about what is going on, ask the questions. If you feel something is being done poorly or incorrectly, bring the subject up for discussion. This is not to say that all things will change or things will always go the way you think they should go, but you will have had input into those things that affect you personally.

Try to see basic training from the perspective of the particular military service. If, for example, we were to look at what the Army hopes to accomplish in basic training, we would see that it has a number of goals. The Army is interested in promptly identifying those men and women who would never be productive in the Army or would never be able to adapt to military life. The Army is also interested in providing realistic training that will be good preparation for the hardships of combat. The Army must present training in military skills to a group of trainees the vast majority of whom have never had any exposure to the military. The Army must do this while instilling the values of esprit de corps, unit pride, and appreciation of Army traditions. And above all, the Army, like the other services, must be sensitive to the thoughts, feelings, and aspirations of the individual. The most important asset in all the services is the individual soldier, sailor, or airman. The services realize this. Training has to be conducted with the individual's welfare in mind.

If you understand what the military services are trying to do, you can appreciate that they have a difficult task. If you can understand their goals and then assist the service in accomplishing those goals, you will get through basic training with a great feeling of accomplishment and self-esteem. You will also have matured greatly.

*Chapter* III

# Pay and Benefits

Both tangible and intangible rewards are available in the armed forces. First we shall look at the tangible rewards and then discuss briefly some of the intangible benefits. Military personnel in all five services are paid according to the same pay scale and receive the same basic benefits. Military pay and benefits are set by Congress, which normally grants a cost-of-living pay increase once each year. In addition to pay, the military provides many of life's necessities, such as food, clothing, and housing, or pays monthly allowances for them. Enlistees can progress through nine enlisted pay grades during their career. Pay grade and length of service determine a servicemember's pay. Table 2 shows the relationship between pay grade and rank.

Recruits begin at pay grade E-1, except that in some services a few who have technical job skills enter at a higher pay grade. Within six months enlistees usually move up to E-2. Within the next six to twelve months, the military promotes enlistees to E-3 if job performance is satisfactory and other requirements are met. Promotions to E-4 and above are based on job performance, leadership ability, and time in the current pay grade. Promotions become more competitive at the higher pay grades.

The major part of an enlistee's paycheck is basic pay. Pay grade and total years of service determine basic pay. Cost-of-living increases generally occur once a year. The military offers incentives and special pay for certain types of duty. For example, incentives are paid for submarine and flight duty and such hazardous duty as parachute jumping, flight deck duty, and explosives demolition. In addition, the military gives special pay for sea duty, diving duty, special assignments, duty in some foreign places, and duty in areas subject to hostile fire. Depending on the service, bonuses are also paid for entering certain occupations.

Most enlisted members, especially in the first year of service,

Figure 2

Insignia of the United States Armed Forces

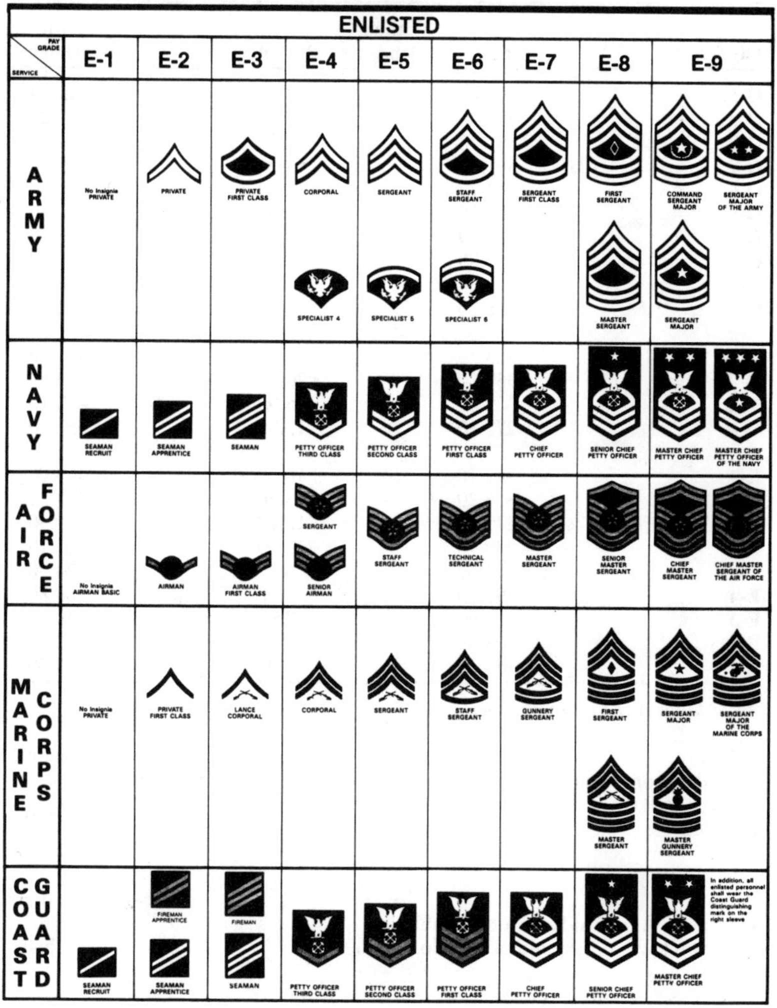

| SERVICE / PAY GRADE | E-1 | E-2 | E-3 | E-4 | E-5 | E-6 | E-7 | E-8 | E-9 | |
|---|---|---|---|---|---|---|---|---|---|---|
| **ENLISTED** | | | | | | | | | | |
| **A R M Y** | No Insignia PRIVATE | PRIVATE | PRIVATE FIRST CLASS | CORPORAL / SPECIALIST 4 | SERGEANT / SPECIALIST 5 | STAFF SERGEANT / SPECIALIST 6 | SERGEANT FIRST CLASS | FIRST SERGEANT / MASTER SERGEANT | COMMAND SERGEANT MAJOR / SERGEANT MAJOR | SERGEANT MAJOR OF THE ARMY |
| **N A V Y** | SEAMAN RECRUIT | SEAMAN APPRENTICE | SEAMAN | PETTY OFFICER THIRD CLASS | PETTY OFFICER SECOND CLASS | PETTY OFFICER FIRST CLASS | CHIEF PETTY OFFICER | SENIOR CHIEF PETTY OFFICER | MASTER CHIEF PETTY OFFICER | MASTER CHIEF PETTY OFFICER OF THE NAVY |
| **A I R   F O R C E** | No Insignia AIRMAN BASIC | AIRMAN | AIRMAN FIRST CLASS | SENIOR AIRMAN / SERGEANT | STAFF SERGEANT | TECHNICAL SERGEANT | MASTER SERGEANT | SENIOR MASTER SERGEANT | CHIEF MASTER SERGEANT | CHIEF MASTER SERGEANT OF THE AIR FORCE |
| **M A R I N E   C O R P S** | No Insignia PRIVATE | PRIVATE FIRST CLASS | LANCE CORPORAL | CORPORAL | SERGEANT | STAFF SERGEANT | GUNNERY SERGEANT | FIRST SERGEANT / MASTER SERGEANT | SERGEANT MAJOR / MASTER GUNNERY SERGEANT | SERGEANT MAJOR OF THE MARINE CORPS |
| **C O A S T   G U A R D** | SEAMAN RECRUIT | FIREMAN APPRENTICE / SEAMAN APPRENTICE | FIREMAN / SEAMAN | PETTY OFFICER THIRD CLASS | PETTY OFFICER SECOND CLASS | PETTY OFFICER FIRST CLASS | CHIEF PETTY OFFICER | SENIOR CHIEF PETTY OFFICER | MASTER CHIEF PETTY OFFICER | In addition, all enlisted personnel shall wear the Coast Guard distinguishing mark on the right sleeve |

*Figure 2*

## MONTHLY PAY

### Monthly Basic Allowance

| Pay Grade | Full | Without Dependents | Partial | With Dependents |
|---|---|---|---|---|
| O-10 | $581.40 | | $50.70 | $715.20 |
| O-9 | 581.40 | | 50.70 | 715.20 |
| O-8 | 581.40 | | 50.70 | 715.20 |
| O-7 | 581.40 | | 50.70 | 715.20 |
| O-6 | 533.70 | | 39.60 | 646.60 |
| O-5 | 503.70 | | 33.00 | 597.60 |
| O-4 | 461.70 | | 26.70 | 546.30 |
| O-3 | 373.80 | | 22.20 | 455.40 |
| O-2 | 301.20 | | 17.70 | 390.60 |
| O-1 | 258.30 | | 13.20 | 350.10 |
| W-4 | 423.30 | | 25.20 | 491.10 |
| W-3 | 357.30 | | 20.70 | 439.50 |
| W-2 | 321.60 | | 15.90 | 410.70 |
| W-1 | 272.10 | | 13.80 | 357.90 |
| E-9 | 341.10 | | 16.60 | 465.00 |
| E-8 | 316.20 | | 15.30 | 433.20 |
| E-7 | 270.00 | | 12.00 | 402.90 |
| E-6 | 239.70 | | 9.90 | 365.70 |
| E-5 | 221.40 | | 8.70 | 324.90 |
| E-4 | 192.30 | | 8.10 | 280.80 |
| E-3 | 186.60 | | 7.80 | 258.30 |
| E-2 | 158.40 | | 7.20 | 258.30 |
| E-1 | 144.30 | | 6.90 | 258.30 |

**Basic Allowance For Subsistence**

**Officers**                                            $114.90 per month
(including commissioned officers, warrants and
aviation cadets)
**Enlisted**

| | E-1 −4 mos. | All others |
|---|---|---|
| When rations in kind are not available | $5.72 | $6.19 |
| When on leave or granted permission to mess separately | 5.06 | 5.48 |
| When assigned under emergency conditions where no government messing is available | 7.58 | 8.19 |

**Table 2  Active Duty Pay Rates Beginning Jan. 1, 1988**

| Pay Grade | Under 2 | 2 | 3 | 4 | 6 | 8 | 10 | 12 | 14 | 16 | 18 | 20 | 22 | 26 |
|---|---|---|---|---|---|---|---|---|---|---|---|---|---|---|
| **COMMISSIONED OFFICERS** | | | | | | | | | | | | | | |
| O-10 | 5485.80 | 5679.00 | 5679.00 | 5679.00 | 5679.00 | 5896.50 | 5896.50 | 6041.70 | 6041.70 | 6041.70 | 6041.70 | 6041.70 | 6041.70 | 6041.70 |
| O-9 | 4862.10 | 4989.30 | 5095.50 | 5095.50 | 5095.50 | 5225.10 | 5225.10 | 5225.10 | 5442.60 | 5896.50 | 5896.50 | 5896.50 | 6041.70 | 6041.70 |
| O-8 | 4403.70 | 4535.40 | 4643.10 | 4643.10 | 4643.10 | 4989.30 | 4989.30 | 5225.10 | 5225.10 | 5442.60 | 5679.00 | 5896.50 | 6041.70 | 6041.70 |
| O-7 | 3659.10 | 3907.80 | 3907.80 | 3907.80 | 4083.00 | 4083.00 | 4319.70 | 4319.70 | 4535.40 | 4989.30 | 5332.50 | 5332.50 | 5332.50 | 5332.50 |
| O-6 | 2712.00 | 2979.90 | 3174.90 | 3174.90 | 3174.90 | 3174.90 | 3174.90 | 3174.90 | 3282.60 | 3801.60 | 3996.00 | 4083.00 | 4319.70 | 4685.10 |
| O-5 | 2169.00 | 2547.00 | 2723.10 | 2723.10 | 2723.10 | 2723.10 | 2805.60 | 2956.20 | 3154.50 | 3390.60 | 3585.00 | 3693.60 | 3822.60 | 3822.60 |
| O-4 | 1828.50 | 2226.60 | 2374.80 | 2374.80 | 2418.90 | 2525.70 | 2697.90 | 2849.70 | 2979.90 | 3110.40 | 3196.50 | 3196.50 | 3196.50 | 3196.50 |
| O-3 | 1699.20 | 1899.60 | 2030.70 | 2247.00 | 2354.40 | 2439.00 | 2571.00 | 2697.90 | 2764.50 | 2764.50 | 2764.50 | 2764.50 | 2764.50 | 2764.50 |
| O-2 | 1481.70 | 1618.20 | 1943.70 | 2009.10 | 2051.40 | 2051.40 | 2051.40 | 2051.40 | 2051.40 | 2051.40 | 2051.40 | 2051.40 | 2051.40 | 2051.40 |
| O-1 | 1286.10 | 1339.20 | 1618.20 | 1618.20 | 1618.20 | 1618.20 | 1618.20 | 1618.20 | 1618.20 | 1618.20 | 1618.20 | 1618.20 | 1618.20 | 1618.20 |
| **COMMISSIONED OFFICERS WITH MORE THAN 4 YEARS ACTIVE DUTY AS ENLISTED OR WARRANT OFFICER** | | | | | | | | | | | | | | |
| O-3E | 0.00 | 0.00 | 0.00 | 2247.00 | 2354.40 | 2439.00 | 2571.00 | 2697.90 | 2805.60 | 2805.60 | 2805.60 | 2805.60 | 2805.60 | 2805.60 |
| O-2E | 0.00 | 0.00 | 0.00 | 2009.10 | 2051.40 | 2116.20 | 2226.20 | 2311.50 | 2374.80 | 2374.80 | 2374.80 | 2374.80 | 2374.80 | 2374.80 |
| O-1E | 0.00 | 0.00 | 0.00 | 1618.20 | 1728.60 | 1792.20 | 1857.00 | 1921.80 | 2009.10 | 2009.10 | 2009.10 | 2009.10 | 2009.10 | 2009.10 |
| **WARRANT OFFICERS** | | | | | | | | | | | | | | |
| W-4 | 1731.00 | 1857.00 | 1857.00 | 1899.60 | 1986.00 | 2073.60 | 2160.60 | 2311.10 | 2418.90 | 2503.80 | 2571.00 | 2653.80 | 2742.60 | 2956.20 |
| W-3 | 1573.20 | 1706.70 | 1706.70 | 1728.60 | 1748.70 | 1876.70 | 1986.00 | 2051.40 | 2116.20 | 2179.20 | 2247.00 | 2334.30 | 2418.90 | 2503.80 |
| W-2 | 1377.90 | 1490.70 | 1490.70 | 1534.20 | 1618.20 | 1706.70 | 1771.50 | 1836.30 | 1899.60 | 1966.20 | 2030.70 | 2094.90 | 2179.20 | 2179.20 |
| W-1 | 1148.10 | 1316.40 | 1316.40 | 1426.20 | 1490.70 | 1554.90 | 1618.20 | 1685.10 | 1748.70 | 1813.80 | 1876.80 | 1943.70 | 1943.70 | 1943.70 |
| **ENLISTED MEMBERS** | | | | | | | | | | | | | | |
| E-9 | 0.00 | 0.00 | 0.00 | 0.00 | 0.00 | 0.00 | 2013.60 | 2059.20 | 2105.70 | 2154.00 | 2202.00 | 2244.90 | 2362.80 | 2592.60 |
| E-8 | 0.00 | 0.00 | 0.00 | 0.00 | 0.00 | 1688.70 | 1737.00 | 1782.60 | 1829.10 | 1877.10 | 1920.60 | 1967.70 | 2083.20 | 2315.40 |
| E-7 | 1179.00 | 1272.60 | 1320.00 | 1365.90 | 1412.70 | 1457.70 | 1504.20 | 1551.00 | 1621.20 | 1667.40 | 1713.90 | 1736.10 | 1852.80 | 2083.20 |
| E-6 | 1014.30 | 1105.50 | 1151.70 | 1200.60 | 1245.30 | 1290.60 | 1338.00 | 1407.00 | 1451.10 | 1497.90 | 1520.70 | 1520.70 | 1520.70 | 1520.70 |
| E-5 | 890.10 | 969.00 | 1015.80 | 1060.20 | 1129.80 | 1175.70 | 1222.50 | 1267.50 | 1290.60 | 1290.60 | 1290.60 | 1290.60 | 1290.60 | 1290.60 |
| E-4 | 830.40 | 876.60 | 928.20 | 1000.20 | 1039.80 | 1039.80 | 1039.80 | 1039.80 | 1039.80 | 1039.80 | 1039.80 | 1039.80 | 1039.80 | 1039.80 |
| E-3 | 782.10 | 825.00 | 858.30 | 892.20 | 892.20 | 892.20 | 892.20 | 892.20 | 892.20 | 892.20 | 892.20 | 892.20 | 892.20 | 892.20 |
| E-2 | 752.70 | 752.70 | 752.70 | 752.70 | 752.70 | 752.70 | 752.70 | 752.70 | 752.70 | 752.70 | 752.70 | 752.70 | 752.70 | 752.70 |
| E-1 | 671.40 | 671.40 | 671.40 | 671.40 | 671.40 | 671.40 | 671.40 | 671.40 | 671.40 | 671.40 | 671.40 | 671.40 | 671.40 | 671.40 |

E-1 with less than 4 months—620.70

live in military housing and eat in military dining facilities free of charge. Those living off base receive quarters (housing) and subsistence (food) allowances in addition to basic pay. Housing allowances vary from $145 to $465 a month depending on pay grade and number of dependents. The food allowance ranges from $165 to $250 per month. Because allowances are not taxed as income, they provide a significant tax saving in addition to their cash value. When added up, housing and food allowances thus represent substantial additions to basic pay.

Military personnel receive substantial benefits in addition to pay and allowances. While in the service enlisted members receive health care, vacation time, legal assistance, recreational programs, educational assistance, and commissary/exchange (military stores) privileges. Families of servicemembers also receive some of these benefits. Table 3 contains a summary of these employment benefits.

The military offers one of the best retirement plans in the country if you are planning to make a career in the military. The amount of retirement pay is graduated depending on your retirement rank and length of service. You may retire at a maximum of 75 percent of your base pay after thirty years of service. Other retirement benefits include medical care and commissary/exchange privileges. Veterans of military service are entitled to certain veterans' bene-

**Table 3    Summary of Employment Benefits for Enlisted Members**

| | |
|---|---|
| **Vacation** | Leave time of thirty days per year |
| **Medical, Dental, and Eye Care** | Full medical, hospitalization, dental, and eye care services for enlistees and most health care costs for family members |
| **Continuing Education** | Voluntary educational programs for undergraduate and graduate degrees, or for single courses, including tuition assistance for programs at colleges and universities |
| **Recreational Programs** | Programs include athletics, entertainment, and hobbies: |
| | Softball, basketball, football, swimming, tennis, golf, weight training, and other sports |
| | Parties, dances, and entertainment |
| | Club facilities, snack bars, game rooms, movie theaters, and lounges |
| | Active hobby and craft clubs, and book and music libraries |
| **Exchange and Commissary Privileges** | Food, goods, and services at military stores are available, generally at lower costs |
| **Legal Assistance** | Many free legal services for help with personal matters |

fits set by Congress and provided by the Veterans Administration (VA). In most cases these include home loans, hospitalization, survivor benefits, educational benefits, disability benefits, and assistance in finding civilian employment.

Servicemembers, regardless of rank or length of service, earn thirty days of leave (vacation) with pay each year. Leave is accrued at the rate of two and one half days per month. During initial training leave is granted only for emergencies verified by the American Red Cross. Anyone entering active duty for thirty-one days or more is automatically insured for $50,000 at a premium of $4 per month under the Serviceman's Group Life Insurance Program. All five of the armed forces encourage their members to further their education. Each service has numerous programs to help defray the high costs of advanced education.

Veterans who serve on active duty for at least 181 consecutive days become eligible to receive certain VA benefits while on active duty. Members who meet the above time in service who are discharged under other than dishonorable conditions, and members who are discharged with fewer than 181 days of continuous active duty because a service-connected disability are also eligible for benefits. Under the new GI education bill, servicemembers may receive a basic benefit of up to $300 a month for 36 months of approved education, to a total of $10,800. To enter the program, servicemembers must contribute $100 a month for the first 12 months of their enlistment; they must serve not less than 20 months of active duty to qualify for the benefits. Members contributing to the GI education bill have 10 years from their date of discharge to use their benefits. The whole area of military benefits, particularly educational benefits, is always subject to change by Congress, so it is wise to check with your recruiter to obtain the latest information.

Intangible benefits resulting from serving in the military are also substantial. Fortunately for the United States, many young men and women have high ideals and a deep sense of patriotism. In my talks with young people just entering the service, I was greatly impressed with their seriousness of purpose and sense of real values as they related to military service. They expressed a desire to contribute to the security of the United States and to render important service to society. It is not a high-salaried job and material advantages they are seeking; rather, it is an opportunity to serve a useful purpose in the world. Serving in the military provides an opportunity to gain job satisfaction and to do something to serve America.

Chapter IV

# The US Army

## *MISSION*

The Army, along with the Air Force, Navy, and Marine Corps, has one fundamental mission: to provide for the security of the United States and for the support of US national and international policies. The ultimate purpose of all military training is to prepare personnel to carry out efficiently and expeditiously their service responsibilities. Title 10, United States Code, Section 3062, states in part:

> It is the intent of Congress to provide an Army that is capable, in conjunction with the other armed forces, of preserving the peace and security...of the United States;...supporting the national policies;...implementing the national objectives;... and overcoming any nations responsible for aggressive acts that imperil the peace and security of the United States.

The fundamental role of the Army, as the nation's land force, is to defeat enemy forces in land combat and to gain control of the land and its people. It is the traditional policy of the United States to maintain active armed forces of a size consistent with the immediate security needs of the nation. In the event of an emergency the armed forces must have the capability of rapid expansion, and therefore military leaders and specialists must be trained during peacetime. The mission of the Army is worldwide; therefore, the orientation of basic training has to incorporate this fact of worldwide responsibility into the instruction. The dynamic nature of international politics as well as the technology explosion with its concomitant improvements in weapons lethality make the Army mission more complex.

## *ENLISTMENT AND INITIAL PROCESSING*

Enlistment in the Army may be for two, three, four, five, or six years. Applicants must be from 17 to 35 years old, American citizens or registered aliens, and in good health. To determine what careers they are best suited for, all applicants must take the Armed Services Vocational Aptitude Battery (ASVAB). The ASVAB is offered at most high schools.

In most cases qualified applicants can be guaranteed their choice of training or duty assignment. There are often combinations of guarantees that are particularly attractive to those who are qualified. For those who wish to be guaranteed a specific school, a particular area of assignment, or both, the Army offers the Delayed Entry Program (DEP). An applicant for the DEP can reserve a school or an assignment choice as much as one year in advance of entry into active duty. Other enlistment programs include the Army Civilian Acquired Skills Program, which gives recognition to skills acquired through civilian training or experience. The program allows enlistees with previous training to be promoted more quickly than they ordinarily would be. The Army also offers enlistment bonuses for certain specialties. Enlistment programs and options vary from time to time. You should discuss all the options in detail with your Army recruiter.

Once all the enlistment procedures are completed and the day arrives for you to leave for basic training, it will be a day of excitement and anticipation. You will leave from the Military Enlistment Processing Station (MEPS) where you underwent your mental, medical, and administrative processing. If you were in the DEP or time has elapsed since your last visit to the MEPS, you will be interviewed to see if there have been any changes in your eligibility. Be sure to inform the interviewer of any medical or police involvement you may have had since your last visit. You will proceed to the Army Reception Battalion. As the name implies, this is where you are received into the training center. Normally you can expect about a three-day stay at the Reception Battalion before picking up your new Army gear and being assigned to a training company.

Before starting basic training, you go through Reception Battalion processing, which helps prepare new soldiers for training and later military life. Reception Battalion processing includes the following:

- Uniform issue and fitting
- Personnel records processing

- Identification (ID) card issue
- Orientation
- Eye and dental checks
- Casual pay
- Mental testing
- Interview
- Haircut (for men)

Orientation covers postal service, legal assistance, medical facilities, recreational facilities/activities, religious activities, leave and pass policies, post exchange facilities, medical care for dependents, financial care of dependents, movement of dependents, privately owned vehicles, visitors, family correspondence, shipment of civilian clothing, pay and allowances, service obligations, allotments, survivors' benefits and Serviceman's Group Life Insurance.

Classes are given in barracks upkeep, physical training, drill (marching), and other subjects that will help you adjust to Army living. You will learn a great deal at the Reception Station about the way the Army does things. Listen carefully to what you are told and respond quickly and efficiently to instructions. After processing at the Reception Battalion, you proceed to basic training. Men and women receive essentially the same initial training, including weapons instruction, but may be trained separately. By regulation, women cannot be assigned to combat, but the Army believes that no matter what their specialty, soldiers should learn the basic combat skills that will give them the confidence and ability to defend themselves.

## OVERVIEW OF BASIC TRAINING

The Army has developed a program for all newly entering men and women that is both very general in scope for those who have no military experience and specific in nature when training soldiers for a designated Military Occupational Specialty (MOS). All enlistees participate in Initial Entry Training (IET), which is designed to provide them with the skills and knowledge to perform the MOS in the first unit of assignment. Initial Entry Training includes Basic Combat Training (BCT), Advanced Individual Training (AIT), and One-Station Unit Training (OSUT). (As you may have noticed, the Army uses a lot of acronyms for the sake of brevity. If you encounter an acronym and are not sure what it means, go back to the first time it was used where it is spelled out. Military publications utilize the same system; the first time a name is used it is spelled out with the acronym beside it.) Basic Combat Training lasts eight weeks; Advanced Individual Training usually lasts nine weeks, although it may last longer depending on the individual skills being taught. One-Station Unit Training means that the initial entry training is conducted at one installation in one unit with the same cadre (instructors) and one program of instruction (POI). The BCT and AIT instruction are integrated to permit early introduction of Military Occupational Specialty–specific training, followed by adequate reinforcement training to insure mastery.

The objective of IET is to develop a disciplined, motivated soldier who is qualified with a weapon, physically conditioned, and drilled in the elements of soldiering. The object of AIT is to provide the soldier with initial skills required to function effectively in the first unit of assignment. Men and women receive the same Basic Combat Training and AIT.

Initial Entry Training is conducted in two ways. All combat and some combat-support soldiers attend One-Station Unit Training at one of the following locations:

- Infantry—Fort Benning, Georgia
- Armor—Fort Knox, Kentucky
- Field Artillery—Fort Sill, Oklahoma
- Air Defense Artillery—Fort Bliss, Texas
- Combat Engineer—Fort Leonard Wood, Missouri
- Military Police—Fort McClellan, Alabama

As mentioned, the One-Unit Station Training provides both Basic Combat Training and Advanced Individual Training. Soldiers in other occupational fields attend BCT for eight weeks, during which they learn common skills, then move on to AIT to learn occupational field skills. Basic Combat Training is given at Fort Dix, New Jersey, Fort Jackson, South Carolina, Fort Knox, Fort Leonard Wood, and Fort Sill.

Where you attend basic training depends on the terms of your enlistment agreement, the branch you chose to enter, and where you enlisted in the Army. However, you will be told where you will take basic training; it will not be a surprise. Again, talk at length with your recruiter so that you are absolutely clear on the terms of your enlistment; this includes location of basic training, location of advanced training, and future assignment possibilities.

## *ARMY TRAINING PHILOSOPHY*

The Army states that you do not need to bring any special skills to basic training. If you have qualified for enlistment, the Army believes that the proper training will enable you to make the transition from civilian to military life. But as was said before, you are much better off if you are prepared for basic training by knowing what to expect, and also if you are in excellent physical condition when you report. Many unrealistic movies and articles about basic training portray the process as one in which the Army "tears you down and then builds you up again starting from scratch." In this version the Army degrades you and treats you with scorn until you prove how tough you are. The drill sergeants are mean and sadistic and are not accountable to anyone for their actions. As an individual trainee, you must be a "lone wolf," relying only on yourself and eventually proving your toughness by having a fight with the drill sergeant or another member of your platoon before everybody finally accepts you. Fortunately for today's military, this representation of basic training is completely false.

Military training, especially Basic Combat Training, respects the dignity and welfare of new soldiers. The Army objective is for the individual soldier to look back upon BCT with the feeling that good and rigorous training was conducted by competent professionals. The two cornerstones upon which the BCT program rests are that all soldiers must be capable of performing in a combat situation and that the new soldier must become a team-oriented individual. Team orientation must begin in BCT because it is so deeply a part of all Army operations, and yet it is not encountered in civilian society at large. In addition to these two cornerstones, the Army is concerned with several key issues as described in the BCT program of instruction.

First is the need for the program to be organized with formal intermediate goals or progressive phases so that the conversion process can be properly structured and both trainer and trainee can be clear on progress being achieved. Next is that training be conducted with as much realism, relevance, and combat fidelity as possible to better meet trainer expectations, but more important, to reduce the strangeness of the battlefield environment, sound, weather, smells, sights, physical hardships, and excitement. The training must be carefully integrated and consistent so that the proper mix of knowledge, skill, and attitude elements are presented as they relate to each other.

It is the aim of the training that every day in BCT be structured so that something new is presented, either initially or for evaluation. This takes the form of performance-oriented training requiring mastery of skills or new and different information. The challenge is to conduct the training so that the learning experience is success-oriented and confidence-building. The Army, as well as the other services, is extremely sensitive to the dignity of the new soldier. From the moment you take the Oath of Enlistment you are a soldier, and you will be addressed as such. Every effort is made to instill in you a sense of identification with the uniform, with the training unit, and with the leaders of that unit. This is not accomplished in an atmosphere of "we/they." Rather, from the start of the training cycle you are in an atmosphere that emphasizes "leader/soldier," and the drill sergeant, committee group trainers, and officers want to be seen as role models to be emulated rather than persons to be feared or avoided.

Leaders of training units continually try to develop self-discipline in their soldiers. Development of self-discipline begins early in the BCT cycle through the total control maintained by the training center cadre over all of your activities. This control is relaxed over time as you demonstrate that you are ready to accept responsibility for your actions. You are given ample opportunity to develop self-discipline.

Current Army doctrine organizes BCT into three distinct phases, each addressing a different segment of the soldierization process. Those phases are:

    I. Orientation and Soldierization (Trooper)
    II. Weapons Training (Gunfighter)
    III. Individual Tactical Training (Trailblazer)

Specifically, Phase I is characterized by the total cadre control of troops, absolute adherence to Army standards, constant supervision, platoon integrity, and the beginning of the soldierization process (transition from civilian to soldier).

Phase II continues the enforcement of standards and heavily emphasizes Basic Rifle Marksmanship (BRM) and physical training. Additionally, Phase II features a transition by the Army from an emphasis on platoon activities to an emphasis on company activities. Generally speaking, a platoon has 30 to 40 trainees, and a training company has three or four platoons. A Phase II test completes this segment.

Phase III concentrates on Individual Tactical Training, field training, increased trainee leadership, increased self-discipline, continued company focus, and the end of Phase III test. All previous training is then coordinated into a tactical field training exercise (FTX), an infiltration course, and a 15-mile road march. Graduation and shipping to Advanced Individual Training complete the cycle. The training cadre evaluates and counsels each trainee regarding the goals and standards of each phase prior to his or her advancement to the next phase.

By using the phasing concept and its inherent goals, the Army provides positive direction for young trainees through immediate short-term objectives. The drill sergeants and training cadre communicate the goals and standards for each phase. Goal-setting is fundamental to sound leadership, and constant feedback is vital to goal achievement. The phase training concept formalizes the soldierization process and defines shorter-term goals for all trainees. It establishes and clarifies goals for each phase and provides more structure to the leadership and counseling program. It also provides a workable format for a common-sense, cumulative approach to support the entire training process. The Army bases the soldierization process on an orderly and sequential method, and hopes to produce disciplined, combat skilled, and physically fit soldiers.

## GENERAL INFORMATION ON INITIAL ENTRY TRAINING

If you need one word to describe your first months in the military, the word would be *busy*. You will feel there are not enough hours in the day to accomplish the things you have to do. Your entire duty day (Monday–Saturday) is highly structured and regimented. You get up about 5 a.m. and go to bed about 9:30 p.m., at least through the eight weeks of Basic Combat Training. You may feel that the Army is demanding too much of you, but remember that everybody else in your platoon probably feels the same way. You will learn to manage your time more effectively; you will learn the value of teamwork in accomplishing your assigned tasks. You will feel tired and rushed for the first several weeks. You will be dealing with information and situations almost totally unfamiliar to you. There will probably be times when you think you made the wrong decision in enlisting in the Army. You will want to blame everybody around you when things go wrong. The important thing to remember is that those or similar feelings are normal in stressful situations. Once you convince yourself that no matter how bad things get you are strong enough to overcome all obstacles, you

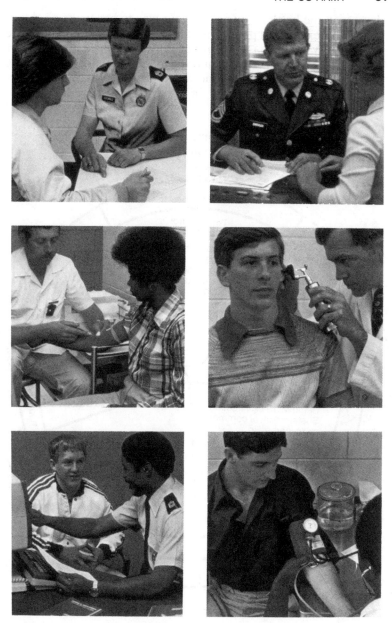

*Enlistees in the Army receive comprehensive physical examinations.*

have won the battle. You are on your way to becoming a soldier.

We spoke earlier of phasing in Basic Combat Training; the entire Initial Entry Training is also broken down into phases. Remember that BCT takes eight weeks, and the advanced individual training usually takes an additional nine weeks. Below is a graphic representation of the five phases of IET. Acronyms in the chart are as follows:

D & C Drill and Ceremonies
BRM Basic Rifle Marksmanship
FTX Field Training Exercise

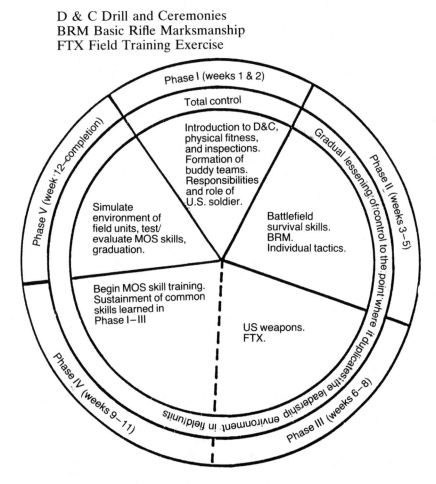

Phase I (weeks 1 & 2)
Total control
Introduction to D&C, physical fitness, and inspections. Formation of buddy teams. Responsibilities and role of U.S. soldier.

Gradual lessening of control to the point where it duplicates the leadership environment in field units.
Phase II (weeks 3–5)
Battlefield survival skills. BRM. Individual tactics.

Phase III (weeks 6–8)
US weapons. FTX.

Phase IV (weeks 9–11)
Begin MOS skill training. Sustainment of common skills learned in Phase I–III

Phase V (week 12–completion)
Simulate environment of field units, test/ evaluate MOS skills, graduation.

*Note:* In OSUT, Phases III and IV may be combined.

*Figure 3*

MOS Military Occupational Specialty
OSUT One Station Unit training

The buddy system (pairing of trainees for mutual assistance and support) is a part of IET. This system helps to reduce and cope with stress, teaches teamwork, and develops a sense of responsibility for fellow soldiers. You are formed into two- to three-person teams upon arrival at the training unit. You learn to help one another in all aspects of training. The system helps in the development of initiative, responsibility, and dependability. When possible, buddy teams participate together in training and other activities. Buddy team changes are limited during IET. Male-female buddy teams are not allowed.

Treatment of trainees follows very strict guidelines. Trainees are treated with the same fairness, respect, and dignity accorded to all soldiers. Degrading of trainees by use of vulgar, obscene, profane, humiliating, or racially or ethnically slanted language is expressly forbidden. Instructors and drill sergeants may touch trainees for the purpose of teaching proper performance. Physical contact with trainees for disciplinary or other reasons is prohibited. Trainees are given a reasonable time to eat their meals. Depriving trainees of meals as a form of discipline is prohibited. Sexual harassment is explicitly forbidden. In BCT training is conducted in all-male and all-female companies. In AIT male and female trainees are integrated for training at the group, class, or platoon level. Trainees are not allowed to wear civilian clothes at any time until they complete Phase III of IET. You are given the option of having the Army store your clothes or sending them home. Proper wearing of all Army uniforms is taught in BCT. It is emphasized that it is a good idea to talk with someone who has recently completed BCT to get his or her impressions of the latest programs and procedures. It is better to have an idea of what is going to happen rather than being surprised by developments.

### BASIC COMBAT TRAINING CURRICULUM AND SCHEDULE

On the following pages the Basic Combat Training course is described in two ways: a course summary, and a weekly master schedule. Not all training programs are exactly alike, but all follow the master schedule.

Basic Combat Training is demanding, but if you are properly prepared you can complete it successfully and experience personal

## BCT COURSE SUMMARY

| *Academic Time* | *Program Hours* |
|---|---|
| A. First Aid | 16 |
| B. Nuclear, Biological, and Chemical Defense | 8 |
| C. M16A1 Rifle Marksmanship | 62 |
| D. Hand Grenades | 8 |
| E. U.S. Weapons | 9 |
| F. Individual Tactical Training | 30 |
| G. Marches and Bivouacs | 16 |
| H. Physical Readiness Training | 50 |
| I. Guard Duty Training | 3 |
| J. Soldier Responsibilities and U.S. Army Heritage and Traditions | 2 |
| K. Identification, Preparation, and Wear of the Uniform | 2 |
| L. Inspections | 16 |
| M. Drill and Ceremonies/Platoon Drill Evaluation | 18 |
| N. Military Customs and Courtesies | 3 |
| O. Basic Military Communications | 4.5 |
| P. Military Justice | 1 |
| Q. Basic Map Reading | 8 |
| R. Code of Conduct | 1 |
| S. Threat Orientation | 1 |
| T. Law of Land Warfare | |
| U. Conditioning Obstacle Course | 4 |
| V. Confidence Obstacle Course | 3 |
| W. Rifle Bayonet Training | 10 |
| X. Hand-to-Hand Combat | 4 |
| Y. Personal Health and Hygience | 1 |
| Z. Field Training Exercise | 19 |
| AA-1. Personal Affairs | 2 |
| AA-2. Alcohol and Drug Abuse Prevention and Control | .5 |
| AA-3. Equal Opportunity | 1 |
| AA-4. Personal Financial Management | .5 |
| Subtotal: | 303.5 |

| *Administrative Time* | |
|---|---|
| Company Commander Time | 4 |
| Reinforcement Time | 34 |
| CIF Issue | 2 |
| Uniform Fitting | 6 |
| Climate Orientation | .5 |
| Commander's Orientation | 1 |
| Immunization | 2 |
| Chaplain's Orientation | .5 |
| Equipment Turn-In | 3 |
| Payday Activities | 4 |
| Maintenance | 28 |
| Movement | 16.5 |
| Inprocessing | 1 |

| | |
|---|---:|
| Outprocessing | 1 |
| Graduation Activities | 4 |
| Guard Duty/Detail | 16 |
| National Holiday | 8 |
| Subtotal: | 131.5 |

*Examination Time*

| | |
|---|---:|
| Phase I Test | 2 |
| Phase II Test | 2 |
| Phase III Test | 5 |
| Subtotal: | 9 |

*Total Course Hours:*     445

## BCT RECOMMENDED MASTER SCHEDULE

| Week: | 1 | 2 | 3 | 4 | 5 | 6 | 7 | 8 | Total |
|---|---|---|---|---|---|---|---|---|---|
| First Aid | | 13 | 2 | | | | 1 | | 16 |
| Nuclear Biological Chemical | | | 8 | | | | | | 8 |
| Individual Tactical Training | | | 3 | | 10 | | 10 | 7 | 30 |
| Marches and Bivouacs | | 4 | | | | 4 | | 8 | 16 |
| Army Physical Readiness Test | 9 | 6.5 | 6 | 6 | 6.5 | 5 | 7 | 4 | 50 |
| Guard Duty | 3 | | | | | | | | 3 |
| Solider Responsibility and U.S. Army Heritage and Traditions | 2 | | | | | | | | 2 |
| Indentification, Preparation, and Wear of Uniform | 2 | | | | | | | | 2 |
| Inspections | 4 | 2 | 2 | 2 | 2 | 4 | | | 16 |
| Drill and Ceremonies | 7 | 4 | 3 | | 4 | | | | 18 |
| Military Customs and Courtesies | 3 | | | | | | | | 3 |
| Military Communications | 4.5 | | | | | | | | 4.5 |
| Military Justice | 1 | | | | | | | | 1 |
| Basic Map Reading | | | 4 | 4 | | | | | 8 |
| Code of Conduct | | | | | | 1 | | | 1 |
| Threat Orientation | | 1 | | | | | | | 1 |
| Law of Land Warfare | | | | | | 1 | | | 1 |
| Rifle Bayonet Training | | 3.5 | | | | | | | 3.5 |
| Total | 31 | 37.5 | 29 | 12 | 22.5 | 15 | 18 | 19 | 184 |
| | | | | | | | | | |
| Pugil Stick | | | 2 | 1.5 | | | | | 3.5 |
| Hand-to-Hand | | | 4 | | | | | | 4 |
| Conditioning Obstacle Course | 2 | | | | | 2 | | | 4 |
| Confidence Obstacle Course | | | 3 | | | | | | 3 |
| Personal Affairs | 2 | | | | | | | | 2 |
| Bayonet Assault Course | | | | | | 3 | | | 3 |
| Sure Pay | | | | | | | | .5 | .5 |
| Alcohol/Drug Abuse | .5 | | | | | | | | .5 |

*An Army helicopter armament specialist readies an AH-64 Apache for flight.*

| | | | | | | | | Total |
|---|---|---|---|---|---|---|---|---|
| Equal Opportunity | 1 | | | | | | | | 1 |
| Personal Health and Hygiene | 1 | | | | | | | | 1 |
| Commander's Orientation | 1 | | | | | | | | 1 |
| Climate Orientation | .5 | | | | | | | | .5 |
| Chaplain's Orientation | .5 | | | | | | | | .5 |
| Field Training Exercise | | | | | | | 9 | 10 | 19 |
| Basic Rifle Marksmemship | 4 | | 4 | 34 | 12 | 6 | | 2 | 62 |
| Hand Grenades | | | | | 8 | | | | 8 |
| U.S. Weapons Testing | | | | | | 9 | | | 9 |
| Phase I Test | | 2 | | | | | | | 2 |
| *Total* | 12.5 | 2 | 13 | 35.5 | 20 | 20 | 9 | 12.5 | 124.5 |

| | | | | | | | | Total |
|---|---|---|---|---|---|---|---|---|
| Phase II Test | | | | | 2 | | | | 2 |
| Phase III Test | | | | | | | 5 | | 5 |
| Central Issue Facility | 2 | | | | | | | | 2 |
| Uniform Fitting | | | | | | 6 | | | 6 |
| Immunization | 2 | | | | | | | | 2 |
| Equipment Turn-In | | | | | | | | 3 | 3 |
| Payday Activities | | | 2 | | | | | 2 | 4 |
| Outprocessing | | | | | | | | 1 | 1 |
| Graduation Activities | | | | | | | | 4 | 4 |
| Company Time | | | | | | | 1 | 3 | 4 |
| Reinforcement Time | 3 | 9 | | | 8 | 5 | 9 | | 34 |
| Inprocessing | 1 | | | | | | | | 1 |
| Weapon/Equipment Maintenance | | | | | | | | 8 | 8 |
| *Total* | 8 | 11 | | | 10 | 11 | 15 | 21 | 76 |

growth. Personal time is limited, but plenty is allowed to receive and answer mail, for personal care, and to attend church. You will successfully complete Basic Combat Training in the Army if *you* want to.

You most assuredly will travel in the military, and the majority of people in the services travel to places outside of the United States. You will meet many new people and form lasting friendships. You will be serving in an honored profession that offers promotions on merit and provides excellent job security. You will have almost unlimited educational opportunities and become part of an organization that cares for you and your family. I do not mean to imply that military service is for everyone; that is certainly not the case. Some people cannot or will not adapt to military life. However, the vast majority of men and women who serve in the military say they are glad they served and feel stronger and more mature as a result of their experiences. For the young man or woman recently graduated from high school, the military services are great places to start.

# Chapter V

# The US Navy

## MISSION

The Navy plays an important role in preserving the national security of the United States. It defends the right of our country and its allies to travel and trade freely on the world's oceans and helps protect our country during times of international conflict. Navy sea power makes it possible for our country to use the oceans when and where our national interests require. The mission of the US Navy is to conduct prompt and sustained combat operations at sea in support of national policy. The principal functions of the mission are sea control, power projection, and strategic sealift. These functions are closely related. A degree of sea control is mandatory in the area from which power is projected, and the ability of naval forces to project power was developed as a method of achieving or maintaining sea control. The Navy's long-standing role of providing strategic sealift is now formally recognized as a third major function. The Navy truly has a worldwide mission. One has only to look at current naval deployments to understand the tremendous scope of that mission. The escorting of US flag ships in the Persian Gulf is just the latest example of employment of US naval forces on a worldwide basis.

## OVERVIEW OF ENTRANCE PROCEDURES

The Navy is a large and diverse organization. It is made up of about 550,000 officers and enlisted people who operate and repair more than 500 ships and over 6,000 aircraft. Navy people serve on ships at sea, on submarines under the sea, in aviation positions on land and sea, and at shore bases around the world. The Navy recruits 80,000 enlisted men and women each year to fill openings in Navy career fields.

To qualify for enlistment in Navy programs, men and women

must be between the ages of 17 and 34. Parental consent is required for 17-year-olds. In the nuclear field, the maximum enlistment age is 23 because of extensive training requirements. Since many Navy programs require enlistees to be high school graduates, the Navy prefers young people to graduate before entering the Navy. Initial enlistment usually is for four years; however, three-, five-, or six-year enlistments are available depending on the program selected. After going through the enlistment process at a Military Entrance Processing Station, you probably would be placed in the Delayed Entry Program (DEP), which guarantees training assignments. The program allows you to finish high school, take care of personal business, or just relax before reporting for duty.

Extra pay is awarded for sea duty, submarine duty, demolition duty, diving duty, service in combat zones, and for work as a crew member of an aviation team or in jobs that require special training. Because the nuclear field is such a critical and unique area of the Navy, quicker promotions are earned and bonuses are available when the training is completed as well as when you enlist.

One of many things that attract people into the Navy is its well-deserved reputation for excellent training. The Navy provides both recruit training and job training. Although recruit training is our main interest here, you should know that job training is a continuous process after you finish recruit training. After recruit training, most Navy people go directly to the technical school (called Class A school) for which they signed up at the Military Entrance Processing Station. Navy Class A Schools are located on military bases throughout the United States, including Great Lakes, Illinois; San Diego, California; Newport, Rhode Island; and Pensacola, Florida. These schools range in length from a few weeks to many months, depending on the complexity of the subject.

## SUMMARY OF RECRUIT TRAINING

Your first assignment as a Navy enlistee is recruit training. It is a tough eight-week period of transition from civilian to Navy life. It provides the discipline, knowledge, and physical conditioning necessary to continue serving in the Navy. Navy recruit training centers are located in Orlando, Florida; Great Lakes, and San Diego. Women recruits train only at Orlando.

After reporting, recruits are placed in training companies, issued uniforms and equipment, and assigned living quarters. The recruit's

*Meteorology is a specialty open to women Navy recruits.*

day starts at 0530 (5:30 a.m.). Taps (lights out) is at 2130 (9:30 p.m.). During weekdays the schedule is based on ten periods of physical fitness and classroom instruction, each lasting 40 minutes.

Physical fitness training includes push-ups, sit-ups, jumping jacks, distance running, water survival, and swimming instruction. Recruits are tested for physical fitness at the beginning and end of recruit training. The test requirements differ for men and women.

Classroom and field instruction covers more than 30 subjects, including aircraft and ship familiarization, basic deck seamanship, career incentives, decision-making, time management, military drill, Navy mission and organization, military customs and courtesies, and the chain of command.

Upon completion of recruit training, seamen proceed to Class A schools or are ordered to the Fleet for duty and on-the-job training through a four-week apprenticeship cruise designed to enhance basic technical skills learned in recruit training. Some recruits are promoted meritoriously to E-2 upon completion of basic training. A Navy seaman recruit (E-1) is advanced to seaman apprentice (E-2) upon completion of six months of active duty, with the commanding officer's recommendation.

*Navy occupations range from the traditional to the high-tech.*

## WHAT TO EXPECT AT RECRUIT TRAINING

Recruit training or basic training conducted by all the armed forces is similar in many ways, but there are also significant differences caused by each service's unique mission. The training philosophy of the Navy, like that of the other services, is to present rigorous, intellectually stimulating, and physically demanding training that enables the recruit to make the transition from civilian to military life, but it also imbues the recruit with an appreciation of naval traditions and heritage.

Recruit training is not easy. The first three weeks are especially tough. You are faced with obstacles to overcome and standards to meet. A positive attitude can make the training more pleasant and help insure your success at the Recruit Training Center (RTC). Do not go into training with the intention to criticize everything you don't like or don't understand—that is a sure-fire recipe for trouble. Rather, if there are things you don't like or understand, talk them over with someone in authority. You will make it through the training if you tell yourself that no matter how tough it is you are going to make it. Be guided completely in all matters by your company commanders (CC), who are chief petty officers and petty officers (enlisted members with higher ranks). Decide that you will do exactly *what they tell you to do, when they tell you to do it.* Obedience to authority is the first necessary attitude to adopt. You should also try to demonstrate cooperation, teamwork, and determination.

Recruits must satisfactorily complete all phases of training. If you do not meet acceptable academic, physical, or medical standards—or your attitude is poor—you could be set back in your training. However, the Navy will give you a second chance to meet the standards. You should have no difficulty proceeding on schedule if you keep up with daily classes and notes, budget your time, and take advantage of study time.

All recruits begin their Navy career at a naval training center (NTC). Travel to the NTC is arranged and paid for by the Navy following your enlistment at a Military Entrance Processing Station (MEPS) or a Naval Reserve activity. The training is the same at all of the three NTCs. Men are assigned for recruit training to Naval Training Center, Orlando, in central Florida; Naval Training Center, Great Lakes, Illinois, on Lake Michigan about 40 miles north of Chicago; or Naval Training Center, San Diego, California, in the San Diego Bay area. Women are trained only at Orlando.

Each naval training center consists of three commands. Each command provides separate services:

Recruit Training Command (RTC) is where you make the transition to military life during your eight weeks of training. RTC gives you a busy schedule of lectures and drills on Navy history, traditions, and customs and regulations, plus instruction in basic military subjects.

Naval Administrative Command (NAC) provides housing, clothing, and pay. NAC also handles recreational and Navy Exchange store (like a civilian department store) facilities; communications, postal, and transportation services; and police and fire protection. All buildings and grounds at the training center are maintained by NAC.

Service School Command (SSC) consists of the schools that provide technical training for various Navy jobs.

Upon arrival at RTC, you join about 80 young men and women who make up your company. Your company will be housed in one building (called a barracks), with each person responsible for the orderliness and cleanliness of the barracks. When you have turned in your orders and obtained your bedding and bunk assignment for the first night, you will feel you have started to become a part of the Navy. You will also fill out many forms including a bedding custody card, receipts for items issued to you, a safe-arrival card for your parents, and a clothing requisition. You will also be required to provide a urine sample, which will be tested for drug use.

You must have a Social Security number and card before you join the Navy. You should memorize your number because many forms you fill out in the Navy will require the number. The day of arrival at RTC is called "receipt day." You will spend your first day or more at RTC getting ready for recruit training. During orientation before training starts, you will learn the basic routine. You will go through preliminary processing and obtain all the items you will need during training. You will be issued a chit book (valued between $150 and $160), with coupons to spend instead of money for purchases at the Navy Exchange. You do not need to take much cash with you to recruit training; the Navy recommends a maximum of $25. The value of the coupons used is deducted from your pay. You will also undergo a complete medical evaluation. If you need dental work, it will be scheduled.

Soon after reporting in, you will be placed in a company. You will meet the men and women you will be with for the next two months and be introduced to your company commanders (CCs). Your company will be assigned a number and presented with a flag guidon bearing that number. The first day of training is called one-one (1–1) day, which means the first week and the first day of training. The rest of the week is known as one-two day, one-three day, one-four day and so on. Each training company is taken through training by its company commanders, who are chief petty officers and petty officers familiar with instructional techniques, principles of leadership, and administrative procedures. A company commander instructs you in military and physical drills and shows you how to keep yourself, your clothing, equipment, and barracks in smart, shipshape condition. Your company commanders are always available to answer questions. They are there to help you, and you should not be hesitant about approaching them for assistance. Follow the example of your company commanders and you will have a good start on a successful Navy career.

During recruit training you are taught how the Navy does things. You must become familiar with salutes, uniforms, customs, ceremonies, Navy routine and time, and Navy terminology. You will spend many hours learning the details of these subjects. You must salute all commissioned officers and warrant officers when addressed by them or when meeting them. While you are in recruit training, you must also salute all company commanders. However, saluting takes place only when wearing a Navy cap or hat. You will encounter military time as soon as you report to RTC. It may seem a bit confusing until you understand the system, but it is very easy to master. All the services use military time.

The Navy runs on a 24-hour day. Hours of the day are numbered from 1 to 24. In the afternoon, instead of starting again with 1, the Navy goes on to 13. The hours, such as 8 a.m., or 7 p.m., are called 0800 (zero eight hundred) and 1900 (nineteen hundred). Never say, "nineteen hundred hours." Hours and minutes in Navy time go like this: 10:45 a.m. is 1045 (ten forty-five), 9:30 p.m. is 2130 (twenty-one thirty).

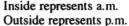

**Inside represents a.m.**
**Outside represents p.m.**

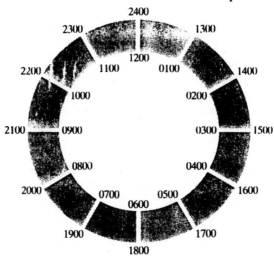

## *NAVAL TERMINOLOGY*

All the military services have their own expressions and jargon. You will learn these terms easily once you use them on a daily basis. Below are some of the terms you will use during recruit training:

| Navy Term | Meaning |
| --- | --- |
| Barracks (divisions) | Building where sailors live |
| Below | Downstairs |

| | |
|---|---|
| Brightwork | Brass or shiny metal |
| Bulkhead | Wall |
| Bunk or rack | Bed |
| Chit, chit book | Coupon or receipt book |
| Chow hall (mess deck) | Place to eat |
| Colors | Raising of a national flag, also, the flag |
| Galley | Kitchen |
| Gear locker | Storage room |
| Geedunk | Candy, gum or cafeteria |
| Head | Bathroom |
| Ladder | Stairs |
| Leave | Authorized vacation |
| Liberty | Permission to leave the base, usually for not more than 48 hours. |
| Overhead | Ceiling |
| Passageway | Hall |
| Rating | A job specialty |
| Reveille | Wake up, start a new day |
| Scullery | Place to wash dishes |
| Scuttlebutt | Drinking fountain or a rumor |
| Secure | Lock, put away, or stop work |
| Sickbay | Hospital or clinic |
| Swab | Mop |
| Taps | Time to sleep, end of day |
| Topside | Upstairs |

## PHYSICAL REQUIREMENTS

Recruits often arrive at RTC completely unprepared for the physical fitness portion on the training. All recruits must meet the physical fitness requirements to graduate from recruit training. Recruits can be set back as a result of failing to meet those requirements—including the swimming qualification. As has been repeatedly emphasized in this book, you are strongly urged to be in top physical shape when reporting to recruit training. It will make everything much easier for you. All recruits take part in daily physical fitness drills. Remedial physical conditioning, with individual instruction, is held for recruits who have trouble with any portion of the program. Remedial swimming instruction is also available for those who cannot pass the swimming qualification.

Physical fitness tests are given during recruit training, and each test is a bit more difficult than the previous one. If you have prepared yourself to pass the first test before reporting to recruit training, you will have no trouble in passing the final test. The test requirements differ for men and women and in some instances may differ slightly at each RTC. Below is a guide to physical fitness standards:

| Men: Requirements | Exercise | First Test | Second Test | Final Test |
|---|---|---|---|---|
| | Push-ups (hands under chin) | 10 | 15 | 25 |
| | Sit-ups (2-count sit-up, knees bent) | 15 | 30 | 45 |
| | Flutter kicks (3 sets of each) | 10 | 15 | 20 |
| | Bodybuilders (8-count) | 10 | 15 | 20 |
| | Jumping jacks | 25 | 50 | 75 |
| | Running | no distance requirement, just run for 10 minutes | 2 miles in 16 minutes | 2 1/4 miles in 18 minutes |

Swimming qualification. Enter water feet first from a height of five feet and float or tread water for five minutes. Then you are required to swim 50 yards.

| Women: Requirements | Exercise | First Test | Second Test | Final Test |
|---|---|---|---|---|
| | Modified push-ups (knee position) | 4 | 15 | 25 |
| | Sit-ups (2-count sit-up, knees bent) | 15 | 30 | 45 |
| | Flutter kicks (3 sets of each) | 10 | 15 | 20 |
| | Mountain climbers | 10 | 15 | 20 |
| | Jumping jacks | 25 | 50 | 75 |
| | Running | just run for 10 minutes | 2 miles in 20 minutes | 2 1/4 miles in 23 minutes |

Swimming qualification. Enter water feet first from a height of five feet and float or tread water for five minutes. Then you are required to swim 50 yards.

## DAILY ROUTINE

There is a set routine for each 24-hour day at recruit training. During weekdays the schedule is based on 10 class sessions daily, each of 40 minutes. Formal classes involve academic instruction, training, and administrative activities. Time scheduled for meals is not counted. The Navy has a reputation for good food, especially on ships at sea. You will find that food served at all recruit training centers is well prepared, nutritious, and plentiful. Here is the approximate daily routine:

*Morning*
| | |
|---|---|
| 0530 | Reveille |
| 0530–0720 | Barracks, clean-up, breakfast |
| 0720–0800 | Training Period 1 |
| 0810–0850 | Training Period 2 |
| 0900–0940 | Training Period 3 |
| 0950–1030 | Training Period 4 |
| 1040–1120 | Training Period 5 |

*Afternoon*
| | |
|---|---|
| 1120–1300 | Lunch |
| 1300–1340 | Training Period 6 |
| 1350–1430 | Training Period 7 |
| 1440–1520 | Training Period 8 |
| 1530–1610 | Training Period 9 |
| 1620–1700 | Training Period 10 |

*Evening*
| | |
|---|---|
| 1700–1830 | Evening meal |
| 1830–1930 | All recruits shower and shine shoes |
| 1930–2015 | All recruits on cleaning stations, clean barracks |
| 2015–2110 | Study period and letter-writing |
| 2110–2125 | Instructions, night bunk check |
| 2125–2130 | Tattoo (preparation for taps) |
| 2130 | Taps (lights out) |

The daily routine at recruit training and elsewhere in the Navy appears in a bulletin called the Plan of the Day (POD). It issues the special orders of the day; gives the hours of meals, inspections, parades, and other events; and names duty officers and duty petty officers. Be sure to check the POD every day. It is usually placed on the bulletin board. The evening routine is an important part of the overall training program and is scheduled to help recruits maintain good performance levels in their studies and other activities required for graduation. During the later weeks of training there may be some relaxation in the evening routine.

The routine for Saturday, Sunday, and holidays is scheduled to continue the training environment of good order and discipline. As approved by your company commander, your weekend time may be spent in organized athletics and competitions, controlled recreation, remedial training, or religious activities. There is usually free time for reading, writing letters, or making phone calls.

## PERSONAL MATTERS

You will receive pay during the third and sixth weeks of training and your final pay balance in the last week. When you arrive you will receive an allowance of between $150 and $160 to help cover expenses. This allowance is in the form of a coupon book called a chit book. Women may receive a larger allowance if they do not bring lingerie and other personal items with them and must purchase them on base.

When you report for training, your parents or next of kin receive notice of your safe arrival and correct address. Visitors are not permitted during training. Pass in review day, during the final week of recruit training, is the first authorized visiting time. You will receive information to send home about this ceremony. You are encouraged to invite your parents to attend the recruit pass in review and to dine with you at the enlisted dining facility on that day. Telephone calls may not be received because of the many recruits in training and the tight schedule; however, you will be allowed to call home soon after your arrival. Time may be scheduled for you to make long distance collect calls during evenings and weekends. Calls may also be authorized as earned privileges or in case of emergency.

Before you leave home for the training center, you should inform your family of the following procedure to follow in case of emergency. Should serious illness or death occur in your immediate family after you report to recruit training, your relative should immediately contact the nearest Red Cross office and provide full details of the situation. Your relative should give the Red Cross your name, rank, Social Security number, and military address. Be sure to leave this information with your closest relatives. The Red Cross will contact you through military channels, if necessary. Advise your parents not to contact you directly by phone, but rather to go through the Red Cross. The Red Cross will verify all the information and assist you in a number of ways.

One of the first military duties you will perform as a recruit is a

*Classroom training in the Navy is among the finest available anywhere.*

security watch. Security means protecting a station against damage by storm or fire and guarding against theft, sabotage, and other such activities. Security watches involve sentry duty, guard duty, fire watches, and barracks watches. Watchstanding duty is formal military duty. Each company provides a 24-hour security and fire watch for its assigned barracks and surrounding area. Recruits normally are assigned these watches on a two-hour basis. Other department and division watches are stood by recruits as directed by the commanding officer.

In every organization, whether at home with your family or aboard ship, someone has to do the housework. At RTC this work is called fleet duties and routine (FD&R). It includes food service duties in the enlisted mess hall, various administrative tasks, swabbing, and other jobs that have to be done to ensure good living conditions for everyone. You will be assigned FD&R during your fifth week of recruit training.

## CURRICULUM

You will receive much classroom instruction at recruit training as well as practical hands-on exercises. A great deal of the course is devoted to seamanship, survival-at-sea techniques, ship structure, and firefighting instruction. Subject areas include:

Advancement Program
Aircraft Familiarization
Basic Deck Seamanship
Career Incentives/Medical Benefits
CBR (Chemical, Biological, Radiological) Warfare/Defense
Chain of Command
Classification
Code of Conduct/Geneva Convention
Cultural Adjustments
Decision-making and Time Management
Damage Control
Educational Benefits
Enlisted Service Record
Financial Responsibility
Firefighting
First Aid
General Orders
Hand Salute and Greetings
History of the Navy
Honors and Ceremonies
Inspections
Leave and Earnings Statement/Pay
Leave, Liberty, and Conduct Ashore
Military Drill
Mishap Prevention

3-M System (Administrative System)
Navy Mission and Organization
Officer Recognition
Operational Security
Personal Hygiene
Physical Conditioning
Rates and Ratings
Security Information
Ship Familiarization
Survival at Sea
Telephone Talkers
Uniform Code of Military Justice
Watch, Quarter, and Station Bill (individual duties)
Watchstanding

The entire staff is interested in your welfare, and they provide a supportive atmosphere. The program is challenging. Chaplains, psychologists, and a legal staff are available to assist and counsel you. The most modern medical and dental facilities are available for routine and emergency treatment. Staff personnel are on hand to help you adjust to the Navy way of life. You are given full opportunity to attend the religious services of your choice. The RTC has a chapel in which a full range of denominational worship services are conducted by chaplains, who also are available for pastoral counseling and religious education. In some instances, recruit choirs are organized and sing at services. Attendance at chapel services is voluntary.

Remember, it is a good idea to talk to as many people as possible about recruit training, serving in the Navy, duty options, educational benefits, and so on. By knowing what is expected of you, you will be much better prepared. Navy life is an exciting life.

Chapter VI

# The US Air Force

## MISSION

The mission of the Air Force is to provide an air arm that is capable, along with the other armed forces, of preserving the peace and security of the United Sates, providing for its defense, supporting the national policies to carry out national objectives, and overcoming any nation responsible for aggressive acts that imperil the peace and security of the United States. In addition, the Air Force also provides major space research and development support for the Department of Defense and assists the National Aeronautics and Space Administration (NASA) in conducting our nation's space program. Teamed with the other armed forces, the Air Force is prepared to fight and win any war if deterrence fails.

The Air Force flies and maintains aircraft, such as long-range bombers, supersonic fighters, Airborne Warning and Control Systems (AWACS) planes, and many others, whenever and wherever necessary, to protect the interests of America and American allies. Over 575,000 disciplined, dedicated, and highly trained officers and airmen make up today's Air Force. Some pilot aircraft—everything from helicopters to the space shuttle. Many others do the jobs that support the Air Force's flying mission; they may work as firefighters, aircraft mechanics, security police, air traffic controllers, or in many other Air Force career fields. The Air Force currently recruits about 60,000 men and women each year to fill openings in hundreds of challenging Air Force careers.

## OVERVIEW OF ENTRANCE PROCEDURES

Applicants for enlistment in the Air Force must be in good health, possess good moral character, and make the minimum scores on the Armed Services Vocational Aptitude Battery (ASVAB) required for all Air force enlistment. Men and women

17 to 27 years old may enlist in the Air Force for four or six years under two basic enlistment options: the Guaranteed Training Enlistment Program (GTEP) and the Guaranteed Aptitude Area Enlistment Program. The GTEP guarantees training and initial assignment in a specific skill. The Aptitude Index (AI) Program guarantees classification into one of four aptitude areas (mechanical, administrative, general, or electronic); specific skills within these aptitude areas are selected during basic training. After choosing one of these programs, applicants may also qualify for the Delayed Enlistment Program (DEP). DEP enlistees become members of the Air Force Inactive Reserve with a delayed date for active duty. They do not participate in military activities or earn pay or benefits while in the DEP. Enlistment under the DEP is a legal, binding contract. The individual agrees to enter active duty on a certain date, and the Air Force agrees to accept him or her (if still qualified) and provide training and initial assignment in the aptitude area or job specified.

The Air Force provides two kinds of training to all enlistees: basic training and job training. All Basic Military Training (BMT) is conducted at Lackland Air Force Base in San Antonio, Texas. BMT teaches enlistees to adjust to military life, both physically and mentally, and promotes pride in being a member of the Air Force. It lasts six weeks and consists of academic instruction, confidence courses, physical conditioning, and marksmanship training. Traineees who enlist with an aptitude area guarantee receive orientation and individual counseling to help choose a job specialty compatible with Air Force needs and with their own aptitudes, education, civilian experience, and desires. After graduation from BMT, recruits receive job training in their chosen specialty.

Most BMT graduates go directly to one of the Air Training Command's Technical Centers for formal, in-residence training. In-residence job training is conducted at Chanute Air Force Base (AFB), Rantoul, Illinois; Keesler AFB, Biloxi, Mississippi; Lackland AFB, San Antonio, Texas; Lowry AFB, Denver, Colorado; Sheppard AFB, Wichita Falls, Texas; Goodfellow AFB, San Angelo, Texas; and several other locations nationwide. In formal classes and practice sessions, airmen learn the basic skills needed for first assignment in their specialty. Some airmen proceed directly to their initial duty station and receive instruction in their skill through on-the-job training.

Air Force training does not end with graduation from basic training or a technical training school. Upon arriving at the first permanent duty station, the airmen begin on-the-job training (OJT). OJT is a two-part program consisting of study and supervised job performance. Airmen enroll in skill-related correspondence courses to gain broad knowledge of their Air Force job, and they study technical orders and directives to learn specific tasks they must perform. They also work daily with their trainers and supervisors, who observe them during hands-on task performance. They are also offered advanced training and supplemental formal courses throughout their time in the Air Force to increase their skills in using specific equipment or techniques.

## WHAT TO EXPECT IN BASIC TRAINING

In contrast to the Army and Navy, basic military training for the Air Force is conducted exclusively at a single location. Lackland Air Force Base on the western outskirts of San Antonio, is known as the Gateway to the Air Force. Basic military training is a serious and highly important part of military life. From the day of your arrival through the rest of your military career, you will be expected to abide by training received in basic training as to your conduct, actions, appearance, and all other aspects of Air Force life. This training will provide you with the fundamental knowledge and skills required of a member of the United States Air Force. The subjects you will study include customs and courtesies of the Air Force, drill and ceremonies, military law, familiarization and use of weapons, the Air Force career system, the Military Code of Conduct, human relations, drug and alcohol abuse, and physical conditioning. All trainees, men and women, must fire an M-16 as part of their training in the familiarization and use of weapons.

You will be expected to apply yourself diligently to all training and classroom activities. Your actions for the entire period of basic training and any subsequent technical training are carefully planned and programmed, and you will have very little free time until your training is completed. Normally you will be at Lackland AFB for about six weeks; however, your stay there could be extended for several reasons: illness, injury, recycle in training because of problems in academic areas or physical conditioning, poor attitude or adaptability rating, or changes in technical training requirements.

On your arrival at San Antonio International Airport, you are transported to the Air Force Military Training Center at Lackland AFB. There you are assigned to a training flight with about 45

*Broadcast and recording technicians in the Air Force produce programs for use at home and overseas.*

other newly assigned airmen, taken to a dining hall for your first Air Force meal, and introduced to your military training instructor (MTI), who will escort you to your dormitory. This is where you will reside until completion of training. The following morning (unless you arrive on a weekend) is normally your first day of training. Other processing actions will follow during the remaining weeks of Air Force basic training. During basic training you are not allowed to use or park a privately owned vehicle (POV) on base. Following completion of basic training, you are provided transportation to your first base of assignment.

## FACILITIES

Lackland has five chapels for the use of all personnel. Worship services are held throughout the week for Protestant, Catholic, Jewish, and Orthodox faiths and several denominational groups. Other services include the American Red Cross, which is open 24 hours a day; Air Force Exchange service; three clubs with bingo, shows, dancing, television, radio, pool, Ping-Pong, reading room, and letter-writing facilities available; hobby shops; mail and postal service; banking facilities; Western Union; three theaters; and a library. Should you or your dependents require emergency financial assistance, you may seek help from the Air Force Aid Society through the personal affairs section of the consolidated base personnel office (CBPO).

## PHYSICAL CONDITIONING

This very important part of your basic training is accomplished through a program of supervised exercise and running. Trainees who are unable to perform the required physical exercises and running are evaluated by a physical conditioning specialist and briefed on their specific weaknesses. Remedial exercises and running are prescribed as appropriate. If satisfactory progress is not made, trainees may be recycled in training until such time as they are able to perform the required evaluations. Running and marching are very much a part of basic training. You should condition yourself accordingly before enlistment to prevent possible foot and leg soreness.

## PERSONAL MATTERS

### Homosexuality

All the services are very strict about homosexuality. Homosexuality is not tolerated in any degree in the Air Force. Participation in a homosexual act, or proposing or attempting to do so, is considered serious misbehavior. Similarly, airmen who have homosexual tendencies or who associate habitually with persons known to be homosexuals do not meet Air Force standards. No distinction is made between duty time and off-duty time; the moral standards of the service must be maintained at all times. It is the general policy to discharge members of the Air Force who fall within the purview of this policy. In certain circumstances trial by court-martial with possible punishment under the Uniform Code of Military Justice may be initiated.

### Pay and Clothing

On the second training day, men are paid $130, and women are paid $230 (women are required to buy more items). This is advance partial pay, and some of it is used to buy items needed during the first few weeks of training (toilet articles, stamps, stationery, etc.). On the fifteenth day of training, you receive another

$100. The remainder is paid to you on your thirtieth training day. You should bring at least $25 with you for any unforeseen expenses; enlistees arriving on Friday night or on a holiday are not paid until their second training day. You receive a partial uniform issue during your first week and are then required to store all other outer civilian clothing. When you go for basic training take only enough clothing for a maximum of three days, but be sure it is adequate and suitable for the season. If you received an initial uniform issue from the Air Force Reserve or Air National Guard within the last 90 days, you will receive only partial issue; therefore, you should bring your uniforms to basic training.

## Leave and Pass Policy

Leaves during basic training and delay en route (leave between completion of basic training and beginning of technical training) are granted only for emergency reasons and must be verified by the American Red Cross. You should instruct your family to contact the nearest office of the American Red Cross and provide full particulars of the emergency and your name, Social Security number, and military address. The Red Cross has trained personnel who can contact the proper officials at your Air Force base so that appropriate action may be taken. Calls to the AF Recruiting Office or to your commander will result only in unnecessary delay and expense. Neither routine leave or delay en route will be granted because of normal pregnancy or childbirth, to get married, to resolve marital problems or threatened divorce, to resolve financial problems, because of psychoneurosis based on family separation, or to settle the estate of a deceased relative.

Following completion of basic training, airmen receiving directed duty assignments may be granted a delay en route when no technical training school is involved; however, such delays en route will be charged to leave that is not yet accrued. You accrue leave at the rate of 2½ days per month or 30 days per year. Accordingly, to prevent potential future hardship, you should use leave only when necessary. If you are assigned to attend technical training school, you will be transferred directly from Lackland AFB to the technical training center. Holiday leaves are permitted during suspension of technical training, which normally occurs as of midnight, 21 December (or the preceding Friday when 21 December falls on a weekend), and resumes 0600, 3 January (or the following

Monday when 3 January falls on a weekend). As stated above, you would be using leave that is not yet accrued, and you may remain at the training center during holidays if you do not want leave. During basic training you normally are authorized a 1-day pass during your last week of training. These passes are for Saturday or Sunday from 9 a.m. (0900) to 8 p.m. (2000). Exceptions may be made in case of emergency. Off-base passes are considered a privilege and may be denied to all or part of a flight if not justified by acceptable training performance.

### Air Force Grooming Standards

Members of the Air Force must maintain high standards of dress and personal appearance. The image of disciplined service men or women requires standardization and uniformity to exclude the extreme, the unusual, and the fad. Guidelines are required for the sake of neatness, cleanliness, safety, and military image. Therefore, uniforms must be kept clean, neat, correct in design and specifications, and in good condition. Shoes are required to be shined and in good repair. For men, hair and sideburns must be neat, clean, trimmed, and present a groomed appearance. Beards or goatees are not permitted except when specifically authorized for medical reasons. Mustaches are not permitted in basic training. You are provided with specific details and guidance on these standards upon arrival at the basic training center. While at Basic Military Training School, women must keep their hair clean and styled to present an attractive appearance. The style must be such that the hair does not touch the lowest portion of the collar or fall below the eyebrows. Accounterments such as pins and barrettes may be used only if they are plain, conservative, and similar in color to the person's hair. Earrings, including posts, are not normally worn with the uniform during basic training; however, posts may be worn at night. Women whose ears are pierced may wear small conservative gold, white pearl, or silver spherical earrings with all uniforms, except when safety considerations dictate otherwise.

### Excessive Weight

If your weight exceeds the allowable maximum you will not be allowed to enlist at the MEPS. In fact, all of the military services have weight standards that you must adhere to the entire time you are in the service. The weight standards are as follows:

**Weight Standards—Male and Female**

| Height in | cm | Men Minimum | Maximum | Women Minimum | Maximum |
|-----------|-----|-------------|---------|---------------|---------|
| 60 | 152.40 | 100 (45.45) | 153  (69.54) | 92 (41.48) | 130 (59.09) |
| 61 | 154.94 | 102 (46.36) | 155  (70.45) | 95 (43.18) | 132 (60.00) |
| 62 | 157.48 | 103 (46.81) | 158  (71.81) | 97 (44.09) | 134 (60.90) |
| 63 | 160.02 | 104 (47.27) | 160  (72.72) | 100 (45.45) | 136 (61.81) |
| 64 | 162.56 | 105 (47.72) | 164  (74.54) | 103 (46.81) | 139 (63.18) |
| 65 | 165.10 | 106 (48.18) | 169  (76.81) | 106 (48.18) | 144 (65.45) |
| 66 | 167.64 | 107 (48.63) | 174  (79.09) | 108 (49.09) | 148 (67.27) |
| 67 | 170.18 | 111 (50.45) | 179  (81.36) | 111 (50.45) | 152 (69.09) |
| 68 | 172.72 | 115 (52.27) | 184  (83.63) | 114 (51.81) | 156 (70.90) |
| 69 | 175.26 | 119 (54.09) | 189  (85.90) | 117 (53.18) | 161 (73.18) |
| 70 | 177.80 | 123 (55.90) | 194  (88.18) | 119 (54.09) | 165 (75.00) |
| 71 | 180.34 | 127 (57.72) | 199  (90.45) | 122 (55.45) | 169 (76.81) |
| 72 | 182.88 | 131 (59.54) | 205  (93.18) | 125 (56.81) | 174 (79.09) |
| 73 | 185.42 | 135 (61.36) | 211  (95.90) | 128 (58.18) | 179 (81.36) |
| 74 | 187.96 | 139 (63.18) | 218  (99.09) | 130 (59.09) | 185 (84.09) |
| 75 | 190.50 | 143 (65.00) | 224 (101.81) | 133 (60.45) | 190 (86.36) |
| 76 | 193.04 | 147 (66.81) | 230 (104.54) | 136 (61.81) | 196 (89.09) |
| 77 | 195.58 | 151 (68.63) | 236 (107.27) | 139 (63.18) | 201 (91.36) |
| 78 | 198.12 | 153 (69.54) | 242 (110.00) | 141 (64.09) | 206 (93.63) |
| 79 | 200.66 | 157 (71.36) | 248 (112.72) | 144 (65.45) | 211 (95.90) |
| 80 | 203.20 | 161 (73.18) | 254 (115.45) | 147 (66.81) | 216 (98.18) |

(**Weight in parentheses in kilograms**)

If your weight increases above the maximum allowable level while you are in basic training, you will be recycled in training until you meet the Air Force standards. Failure to meet weight standards could lead to discharge from the service.

### Medication

If you are taking medication that was prescribed for you by a doctor, take it with the prescription to basic training. Female applicants who are taking birth control pills are encouraged to continue to take the medication. They should take their supply of pills to basic training even if the container does not bear a pharmacy label; they will be given refills or a new prescription.

### Mail

Your specific military mailing address will be determined at Lackland. It is recommended that you do not request third class mail (newspapers, magazines, etc.) to be forwarded to you during basic training, as your stay will be relatively short and another change of address will be needed when you depart. You should also discourage relatives and friends from sending packages that contain food, as food is prohibited in the dormitories.

## CHAPEL

The chapel program is extensive and includes major faith groups and various denominational services each week. A chaplain will give you a thorough orientation briefing on all aspects of the religious programs on base and throughout the Air Force, plus a special focus on your personal adjustment to Air Force life. You will also be given the opportunity to discuss with the chaplain any personal problems you might have. The Air Force has traditionally made a conscious effort to provide for the religious practices of all its members.

In this regard, each applicant should understand that special religious observance, diet, and other such practices may conflict with training schedules or essential operating procedures. If a conflict should arise, military training generally takes precedence because basic military training is a very rhythmic and controlled program. Major conflicts are resolved on an individual basis. Conflicts of this nature are rare after completion of the formal training phase of your enlistment.

## WHAT TO BRING AND WHAT NOT TO BRING

Bring all prescription eyewear (prescription glasses or contact lenses) with you. Contact lenses may not be worn during basic training. You will be issued prescription glasses that must be worn while in basic training.

The following list shows the items enlistees are required to have during basic training. If you own suitable items you may bring them with you.

You should bring sufficient personal hygiene items to last for three days. It is recommended that you limit yourself to essential items, since you can purchase any items desired at reasonable prices after arrival. Your military training instructor (MTI) will determine whether items brought from home conform to the center's standards. Because space is limited, the items you bring should be small or medium in size. The MTI normally arranges for

*Technical specialties are popular among Air Force recruits.*

## Item Required during Basic Training

| Male | | Female | |
|---|---|---|---|
| *Mandatory Items* | | *Mandatory Items* | |
| *Flashlight w/reflective tape ($2.15 aprx) | 1 each | *Flashlight w/reflective tape ($2.15 aprx) | 1 each |
| *Padlock (pintumbler) ($4.30) | 1 each | *Padlock (pintumbler) ($4.30) | 1 each |
| *Batteries ($.45 each) | 2 each | *Batteries ($.45 each) | 2 each |
| Shoe polish (black) | 1 each | Shoe polish (black) | 1 each |
| Marking kit | 1 each | Marking kit | 1 each |
| Shoebrush | 1 each | Shoebrush | 1 each |
| Shine cloth | 1 each | Shine cloth | 1 each |
| Shower clogs | 1 pair | Shower clogs | 1 pair |
| Soap | 1 each | Soap | 1 each |
| **Soap tray | 1 each | **Soap tray | 1 each |
| Toothbrush | 1 each | Toothbrush | 1 each |
| Toothbrush tray | 1 each | Toothbrush tray | 1 each |
| Toothpaste or powder | 1 each | Toothpaste or powder | 1 each |
| Notebook | 1 each | Notebook | 1 each |
| Ballpoint pen (dark blue or black ink) | 1 each | Laundry soap | 1 box |
| Laundry soap | 1 box | ***Bras | 6 each |
| Zipper briefcase | 1 each | ***Underpants | 6 each |
| Razor (may be electric) | 1 each | ***Shoes—pump, black (calf or patent) | 1 pair |
| Blades (when safety razor is used) | 1 package | Stockings (nylons/panty hose) | 6 pair |
| | | Hairnet | 1 each |

Zipper briefcase .................. 1 each
Ballpoint pen
  (dark blue or black ink) ...... 1 each
Laundry bag ..................... 1 each
Deodorant ....................... 1 each
Sanitary napkins/tampons ...... 1 box
Shampoo ......................... 1 each

*Recommended Items*

Stationery ........................ 1 each
Sewing kit ........................ 1 each
Towels ............................ 2 each
Foot powder ...................... 1 each
Cotton balls ...................... 1 box
Lipstick or Chapstick ............ 1 each
Pressing cloth (man's handkerchief) 1 each
Bathrobe .......................... 1 each
Pajamas/nightgown ............... 1 each
Hairbrush ......................... 1 each
Nail file .......................... 1 each
Razor (may be electric) .......... 1 each
Eyeglasses guards ................ 1 each
Glass or cup ...................... 1 each
Shoetrees ..................... As required
PC shirt (with squadron color and number) 1 each
Washcloth ........................ 1 each
Iron .............................. 1 each
Spray starch/fabric finish ....... 1 can

---

Shaving cream
  (when safety razor is used)
  or shaving powder .............. 1 can
Laundry bag ..................... 1 each
Deodorant ....................... 1 each

*Recommended Items*

Shampoo .......................... 1 each
Stationery ........................ 1 each
Sewing kit ........................ 1 each
Towels (if student doesn't have
  sufficient number) ............. 2 each
Foot powder ...................... 1 each
Cotton balls ...................... 1 box
*Utility kit
  (nails clippers and file) ...... 1 each
Aftershave lotion ................ 1 each
Iron .............................. 1 each
Spray starch/fabric finish ....... 1 can
PC shirt (with squadron color
  and number) .................... 1 each

*These items have been picked up and delivered by your instructor and will be paid for by you during your initial BX visit.

**Required if bar soap is used.

***You will receive extra money during training for the purpose of purchasing these items either at BMTS or at your next duty station. We recommend you bring running shoes that are in good repair.

pre-issue from the base exchange (BX) of a pintumbler type padlock and a flashlight. You will be required to pay for these items on your first visit to the BX.

Male enlistees are advised to dress casually and comfortably. Slacks, sportshirt, and jacket or sweater (depending on the season) are recommended. You should also wear comfortable shoes or sneakers. Bring at least one change of underwear and socks and running shoes that are in good repair. Female enlistees are advised to dress modestly, casually, and comfortably. Wear a pantsuit or slacks or jeans with a blouse or sweater, a light raincoat, and comfortable, low-heeled walking shoes or tennis shoes (not platforms, high heels, or sandals). An extra change of clothing is necessary, as it could be four days before you are issued your Air Force uniform. You should also bring a pair of running shoes in good shape.

You should not take expensive jewelry or sports equipment to the training center. Any medicine you have should be accompanied by a prescription from a medical doctor. Any weapons or dangerous instruments found in your possession or belongings will be confiscated and disposed of by Air Force personnel.

## AIR FORCE TERMINOLOGY

You should know that all the services have their own individual expressions and abbreviations. It will not take long to learn the meanings; below are some of the words and phrases you will hear in basic:

### Air Force Term Meaning

| | |
|---|---|
| AI | Aptitude Index — an enlistment option with job classification during BMT. |
| Barracks | Dorm or dormitory where you live. |
| Blouse | Coat, usually the dress uniform. |
| BMT | Basic military training. |
| Boondocks | Woods or wilds, training area. |
| Bunk or rack | Bed. |
| BX | Base exchange, a store where you may buy everything from soap to shoes. |
| Colors | A national flag. |
| Cover | Hat. |
| Drill | To march, usually in a group. |
| Dining Hall/Chow Hall | Where you eat your meals. |

| | |
|---|---|
| Dorm/Dormitory | Building where you live. |
| Dispensary | A place for medical treatment (usually an annex to a hospital). |
| Esprit de corps | Spirit of camaraderie (in service to our country). |
| Flight | A unit of Air Force personnel; in BMT, approximately 47 persons. |
| GI | Government issue, members of all branches of the military are referred to as "GIs." |
| GTEP | Guaranteed Training Enlistment Program; an enlistment option that guarantees your job before enlistment. |
| Head | Bathroom |
| Latrine | Bathroom |
| Leave | Authorized vacation (30 days per year). |
| MTI | Military training instructor. |
| NCO | Noncommissioned officer. |
| NCOIC | NCO in Charge. |
| OIC | Officer in Charge. |
| Police up the area | Straighten up, make neat. |
| Quarters | A place to live, house, dorm, etc. |
| Rainbow | New basic trainee (because of multicolored attire of new arrivals). |
| Reveille | First assembly of the day, usually signaled by a bugle call. |
| Secure | Lock, close, put away. |
| Sick Call | A time to report to the dispensary for examination or treatment. |
| SP | Security policeman. This term is used for both security specialists and law enforcement specialists. |
| Squadron | An Air Force unit composed of at least two flights. |

## DAILY ROUTINE AND CURRICULUM

The thirty days of training, excluding Saturdays and Sundays, are extremely busy for the enlistees. You are not allowed any liberty from your training area until after your 15th day of training, and you are not allowed any liberty off-base until after your 25th day of training. Basic training is a large-scale operation at Lackland. Approximately 200 to 400 men and women arrive daily to begin their military careers. During the first seven days of training, the trainees are kept busy with job classification interviews, uniform fittings, testing, and other activities. You start your personnel records, which follow you throughout your career. You take tests, have medical and dental exams, and receive detailed orientation briefings. By the seventh day of training, classes begin. Your 12-hour days are full, consisting of academics, military training, and physical conditioning. Academic instruction includes Air Force organization and history, career advancement, financial management, customs and courtesies, human relations, group living, teamwork, illegal use of drugs, first aid, personal affairs and security, and various other subjects. In addition, all trainees go through the confidence course and fire the M-16 rifle.

The following is a broad outline of your daily schedule. It is a typical winter schedule; in the summer, physical conditioning is scheduled for the early morning.

| | |
|---|---|
| 0400–0500 | Military Training Instructors report for duty |
| 0500 | Duty day begins for trainees |
| Morning activities | Breakfast |
| | Processing |
| | Academic classes |
| | Military instruction |
| 1230 Afternoon activities | Lunch |
| | Parade |
| | Retreat |
| | Confidence course |
| | Marksmanship |
| | Physical conditioning |
| 1730 Evening activities | Dinner |
| | Military training/mail call |
| | Study |
| | Personal time |
| 2100 | Lights out |

The training schedule will give you a good idea of the activities and classes you will be involved in at basic military training. There may be minor changes in each training cycle. (DOT = Day of Training.)

## Male and Female Master Training Schedule

**1 DOT**
3 —Meals
1 —Dorm Arrangement
3 —Initial Clothing Issue
1.5 —Initial BX Visit
3 —24-Hour Orientation
1 —Buddy Briefing
1 —Pseudofolliculitus
1 —Military Training Time/Fire Drill
.5 —Pay
1 —Clipper Cuts

**2 DOT**
3.5 —Meals/Reveille
3 —Medical Dental Proc.
1 —Immunizations/TB Time Test
2 —AFMET Testing/G.I. Bill
1 —Reporting Procedures
1 —Clothing Fit Insp.
2 —UCMJ Brg
2 —Dorm Arrangements
1 —Dorm Maintenance
1 —Individual Drill
.5 —Personal Time

**3 DOT**
3.5 —Meals/Reveille
3 —Aptitude Testing
2 —Records Processing
2 —72-Hr Orientation
2 —Individual Drill

**6 DOT**
3.5 —Meals/Reveille
1 —1st Sgt Brg
4 —Interview Session
4 —ASVAB Testing
1 —Mail Call/MTI Brg
1 —Dorm Maintenance
1 —Study Time
.5 —Personal Time

**7 DOT**
3.5 —Meals/Reveille
4 —Blood Donors
1 —Intro to Classroom Procedures
2 —Military Law
1 —G.I. Bill Brg
1 —Military Training Time
1 —Mail Call/MTI Brg
1 —Study Time
1 —Dorm Maintenance
.5 —Personal Time

**8 DOT**
3.5 —Meals/Reveille
1 —Physical Fit/Rubella Rubeola Immunizations
2 —Flight Drill
1 —Drug and Alcohol

**11 DOT**
3.5 —Meals/Reveille
1 —Physical Fitness
2 —Flight Drill
3 —Career Advancement
2 —Human Relations
1 —Mail Call/MTI Brg
1 —Study Time
.5 —Personal Time
1 —Dorm Maintenance

**12 DOT**
3.5 —Meals/Reveille
1 —Physical Fitness
2 —Flight Drill
1 —Fitness Nutrition
4 —AF History & Org.
1 —Dorm Maintenance
1 —Military Training Time
1 —Mail Call/MTI Brg
1 —Study Time
.5 —Personal Time

**13 DOT**
3.5 —Meals/Reveille
1 —Physical Fitness
2 —Flight Drill
2 —AF Security
1 —Open Ranks Insp

2   —Initial Marking
1   —Dorm Arrangements
1   —Dorm Arrangements
.5  —Patio Orientation

4 DOT
3.5 —Meals/Reveille
6   —Career Guidance (3 ANG, AFR Liaison Bfg)
2   —Physical Fitness Orientation, (Prior to 5th DOT)
2   —Formal Dorm Guard Briefing (NLT 5TH)
1   —Immunizations/X-RAY
1.5 —Reading Test

5 DOT
3.5 —Meals/Reveille
1   —Physical Fitness
.5  —Blood Donors Bfg
.5  —BMTS/CC Orientation
1.5 —Chaplain Orientation
.5  —Dental Hygiene Bfg
1   —Sqdn/CC Briefing
2   —Sewing (Name/USAF)
1   —First Week Bfg
1   —Dorm Arrangement
.5  —Mail Call/MTI Bfg
2   —Patio Visit

2   —AF Customs/Courtesies
1   —Military Trng Time
2   —Military Trng Time
1.5 —Mail Call/MTI Bfg
1   —Dorm Arrangement
.5  —Personal Time
.5  —Grade Insignia

9 DOT
3.5 —Meals/Reveille
1   —Physical Fitness
8   —DD Form 398 Check
1   —Mail Call/MTI Bfg
1   —Study Time
.5  —Personal Time
1   —Dorm Maintenance

10 DOT
3.5 —Meals/Reveille
1   —Physical Fitness
2   —Flight Drill
4   —Assessments
1   —Standby Inspection
1   —Military Trng Time
1   —Mail Call/MTI Bfg
1   —Study Time
.5  —Personal Time
1   —Dorm Maintenance

(Wear of the Uniform)
2   —Moral Leadership
1   —Military Trng Time
1   —Military Trng Time
1   —Mail Call/MTI Bfg
1   —Study Time
.5  —Personal Time
1   —Dorm Maintenance

14 DOT
3.5 —Meals/Reveille
1   —Physical Fitness
3   —2nd Clothing Issue
1   —Final Marking
1   —Dorm Arrangement
2   —Military Trng Time
1   —Mail Call/MTI Bfg
1   —Study Time
1.5 —Personal Time
1   —Clothing Fit Insp

15 DOT
3.5 —Meals/Reveille
1   —Physical Fitness
2   —Flight Drill
1   —Pay/Allotment Proc
1   —ID Card Processing
1   —ID/Base Lib Bfg
1   —Clipper Cuts (M)
1   —MTT (F)
1   —Peer Rating
1   —Weight Check
.5  —Military Trng Time
1   —Mail Call/MTI Bfg

1 —Individual Drill

SAT
3 —Meals
2 —Dorm Maintenance
8 —Military Trng Time
1 —Study Time
2 —Personal Time

SUN
3 —Meals
2 —Dorm Maintenance
2 —Religious Activities
4 —Military Trng Time
1 —Study Time
3 —Personal Time

SAT
3 —Meals
1 —Physical Fitness
5 —Military Trng Time
1 —Study Time
2 —Personal Time
2 —Individual Drill
2 —Dorm Maintenance

SUN
3 —Meals
2 —Religious Activities
5 —Military Trng Time
1 —Study Time
2 —Personal Time
2 —Dorm Maintenance

SAT
3 —Meals
2 —Dorm Maintenance
8 —Military Trng Time
1 —Study Time
2 —Personal Time

SUN
3 —Meals
2 —Dorm Maintenance
2 —Religious Activities
5 —Military Trng Time
1 —Study Time
2 —Personal Time

SAT
3 —Meals
10 —Squadron Details
1 —Study Time
2 —Personal Time

SUN
3 —Meals
2 —Religious Activities
5 —Military Trng Time
1 —Study Time
2 —Personal Time
2 —Dorm Maintenance

1 —Study Time
1 —Personal Time

SAT
3 —Meals
2 —Dorm Maintenance
9 —Military Trng Time
2 —Personal Time

SUN
3 —Meals
2 —Dorm Maintenance
2 —Religious Activities
6 —Military Trng Time
2 —Personal Time

SAT
3 —Meals
8 —Student Details
2 —Military Trng Time
1 —Study Time

SUN
3 —Meals
2 —Religious Activities
4 —Military Trng Time
1 —Study Time
3 —Personal Time
2 —Dorm Maintenance

16 DOT
3 —Meals
12 —Student Details/KP
1 —Mail Call/MTI Bfg

17 DOT
3.5 —Meals/Reveille
1 —Physical Fitness
2 —Flight Drill
3 —Rights/Freedoms/Resp
1 —Chemical Warfare
1 —Dorm Maintenance
2 —Military Trng Time
1 —Mail Call/MTI Brg
1 —Study Time
.5 —Personal Time

18 DOT
3.5 —Meals/Reveille
1 —Physical Fitness
1 —Alterations P/U
1 —Clothing Fit Insp
2 —Hometown News Release

21 DOT
3.5 —Meals/Reveille
1 —Physical Fitness
3 —Parade Practice
1 —Realterations Pick-Up
1 —Clothing Fit Insp
2 —AF Customs & Crts
1 —Dorm Maintenance
1 —Mail Call/MTI Brg
1 —Study Time
.5 —Personal Time
1 —Assignment Policies

22 DOT
3.5 —Meals/Reveille
1 —Physical Fitness
2 —Sq Drill (Parade)
2 —Personal Affairs
1 —Open Ranks Insp
      (Wear of Uniform)
2 —Moral Leadership
1 —Dorm Maintenance
1 —Mail Call/MTI Brg
1 —Study Time
1.5 —Personal Time

23 DOT
3 —Meals
12 —Student Details/Dining
      Hall Attendants
1 —Mail Call/MTI Brg

26 DOT
3.5 —Meals/Reveille
1 —Physical Fitness
1 —Clipper Cuts (M)
      MTT (F)
1 —Open Ranks Insp
      (Wear of Uniform)
1 —Sq Drill (Retreat Prac)
1 —Sq Drill (Retreat Cere)
3 —Military Trng Time
1 —Mail Call/MTI Brg
2.5 —Personal Time
* —MAS Brg
1 —Dorm Maintenance

27 DOT
3.5 —Meals/Reveille
1 —Physical Fitness
2 —Written Evals
1 —Test Critique
1 —Dorm Arrangements
3 —Military Trng Time
1 —Mail Call/MTI Brg
1.5 —Personal Time
1 —Drill Evals
1 —Reporting Evals

28 DOT
3.5 —Meals/Reveille
1 —Physical Fitness
2 —Orders P/U DDA Tech School Brg
2 —Military Trng Time
1 —Mail Call/MTI Brg

1 —Flt/Individual Pict
2 —Military Trng Time
1 —Mail Call/MTI Brg
.5—Resource Protection
.5—Confidence Crse Brg
1 —Study Time
.5—Personal Time
1 —Dorm Maintenance

19 DOT
3 —Meals
4 —Confidence Crse
5 —Premarksmanship Trng
.5—Military Trng Time
1 —Mail Call/MTI Brg
1 —Study Time
1.5—Personal Time

20 DOT
3.5—Meals/Reveille
5 —Marksmanship Trng
1 —Dorm Maintenance
1 —Mail Call/MTI Brg
1 —Study Time
1.5—Personal Time
1 —Standby Insp
2 —Self Aid/Buddy Care

24 DOT
3.5—Meals/Reveille
1 —Physical Fitness
2 —Sq Drill (Parade)
3 —Personal Affairs
1 —Dorm Maintenance
1 —Military Trng Time
1 —Mail Call/MTI Brg
1 —Study Time
2.5—Personal Time

25 DOT
3.5—Meals/Reveille
1 —Physical Fitness
3 —Parade
2 —Personal Affairs
1 —Town Pass Brg
1 —Dorm Maintenance
2 —Clothing Fit Insp (Last Look)
1 —Mail Call/MTI Brg
1 —Study Time
.5—Personal Time
* —Traffic Safety Brg

2.5—Personal Time
3 —Parade
1 —Dorm Maintenance

29 DOT
3.5—Meals/Reveille
1 —Physical Fitness
1.5—Written Re Evals
2 —Departure Orien
1 —Sq Drill (Retreat Prac)
1 —Sq Drill (Retreat Cere)
1 —Dorm Maintenance
1 —Military Trng Time
1 —Mail Call/MTI Brg
2 —Personal Time
1 —Weight Check

30 DOT
3.5—Meals/Reveille
1 —Phys Fitness Re Evals
3 —Pay & Travel Arrg
1 —Immunizations
1 —Shipping Brg
2 —Dorm Dep Preparation
1 —Sq Clearance
1 —Mail Call/MTI Brg
2.5—Personal Time

Service in the Air Force means some of the finest technical training in the world. The Air Force, like the other services, is an excellent place to gain valuable work experience. It is an exciting and rewarding experience. By knowing what you are getting into and paying attention to what goes on at basic training, you will complete the program successfully.

Chapter **VII**

# The US Marine Corps

The Marine Corps (pronounced *core*) is one of four separate military services within the Department of Defense. The Marine Corps and the Navy are separate services within the Department of the Navy. The mission of the Marine Corps is as follows:

Be organized, trained, and equipped to provide Fleet Marine Forces for service with the US Fleet in the seizure or defense of advanced Naval bases and for the conduct of such land operations essential to the prosecution of a naval campaign.

Provide detachments for service on armed vessels of the US Navy and security detachments for the protection of property of Naval Stations and bases.

In connection with the Army, Navy, and Air Force, develop the tactical techniques and equipment employed by landing forces in amphibious operations.

Train and equip Marine forces for airborne operations in coordination with the Army, Navy, and Air Force.

Maintain four fully equipped and manned Division/Wings. Marine Regular forces shall comprise three of these Division/Wings. The fourth shall be the Marine Reserve.

Perform such other duties as the President may direct.

To perform this unique mission, 200,000 officers and enlisted Marines in the Corps fly planes and helicopters; operate radar equipment; drive armored vehicles; gather intelligence; survey and map territory; maintain and repair radios, computers, jeeps, trucks, tanks, and aircraft; and perform hundreds of other challenging jobs. Each year the Marine Corps recruits about 50,000 young men and women to fill openings in numerous career fields.

## *OVERVIEW OF ENLISTMENT PROCESS AND INITIAL TRAINING*

Marine Corps enlistment terms are for three, four, or six years depending on the enlistment program. Young men and women enlisting in the Marine Crops must meet exacting physical, mental, and moral standards. Applicants must be between the ages of 17 and 29, American citizens or registered aliens, and in good health. The Armed Services Vocational Aptitude Battery (ASVAB) is used by the Marine Corps to assess each person's vocational aptitudes and academic abilities. The Marine recruiter can arrange for you to take the ASVAB. Applicants for enlistment can be guaranteed training and duty assignment under a wide variety of options, depending on their degree of education and qualifications. Women are eligible to enlist in all occupational specialties except combat arms—infantry, artillery, and tank and amphibian tractor crew members—and some of the combat support and aviation operations specialties. In addition to regular enlistment, the Marine Corps offers two special enlistment programs, the Delayed Enlistment Program and the Enlistment Options Program, which provide great flexibility for applicants.

### *Delayed Enlistment Program*

Students who wish to complete the Marine Corps enlistment process before graduating from high school or a community college may enlist in the Marine Corps Delayed Entry Program (DEP). Enlistment in the DEP allows applicants to postpone their active duty training for up to six months (in special cases, up to a full year). Enlisting in the DEP has two principal benefits: You can finish high school or community college, and highly desirable enlistment programs such as all computer specialties and many aviation specialties can be reserved early.

### *Enlistment Options Program*

The Enlistment Options Program guarantees well-qualified applicants, before they enlist, assignment to one of several military occupational specialties (MOS) in an MOS cluster. The MOS clusters contain every job available in the Marine Corps, ranging from combat arms to motor transport to high technology avionics, electronics, and computer science. Some enlistment options feature cash bonuses as well as formal training programs.

*Training*

Marine Corps training occurs in two sections: recruit training and job training. Upon completing the enlistment process, all applicants enter Marine Corps Recruit Training. Young men undergo recruit training either at Parris Island, South Carolina, or in San Diego, California. All young women attend recruit training at Parris Island. Recruit training is rigorous, demanding, and challenging. The overall goal is to instill in recruits the military skills, knowledge, discipline, pride, and self-confidence necessary to perform as a United States Marine. In the first several days at the recruit depot, a recruit is assigned to a platoon, receives a basic issue of uniforms and equipment, is given an additional physical exam, and takes further assignment classification tests. Each platoon is led by a team of three Marine drill instructors. A typical training day at a recruit depot begins with reveille at 0530; continues with drill, physical training, and several classes in weapons and conduct; and ends with taps at 2130.

Upon graduation from recruit training, each Marine takes a short period of leave, then reports either to a new command for formal schooling or to on-the-job training to which he or she has been assigned. All in all, the Marine Corps sends students to over 200 basic formal schools and to over 300 advanced formal schools. The length of formal school varies from four weeks to over a year, depending on the level of technical expertise and knowledge required to become proficient in certain job skills. For example, different MOSs in the electrical and electronic repair field require from 10 to 50 weeks of training; different MOSs in the vehicle and machinery mechanic field require from 6 to 18 weeks. Job training environments vary depending on the nature of instruction. Marines assigned to an MOS within the combat specialty occupational field receive most of their training in the countryside. Marines receiving training in highly technical MOSs spend most of their time in a classroom. The main thrust of Marine Corps training is toward hands-on learning and practical application of newly acquired skills. As soon as possible after classroom instruction is completed, students are placed in an actual work environment to gain practical experience and to develop confidence. Upon completion of entry-level MOS training, most Marines are assigned to operational units of the Fleet Marine Forces to apply their skills. Marines assigned to the more technical MOSs may require more advanced training before their first operational duty assignment. Job performance requirements in a number of MOSs are comparable to those necessary for journeyman certification in civilian occupations. Marines assigned to these MOSs may apply for status as a registered apprentice. Journeyman status can be earned in Marine specialties such as air traffic control, electricity, and surveying, to name a few.

## WHAT TO EXPECT IN RECRUIT TRAINING

Boot camp is the most physically demanding recruit training of all the military services, for two main reasons. Boot camp or basic training is longer in the Marine Corps (eleven weeks) than in any other service, and the Marine Corps puts more emphasis on physical conditioning and exercise than does any other service. It is extremely important to get yourself in good physical condition before reporting to boot camp. If you are not in good shape, the course will be too demanding for you. You are advised to participate in a progressive program of running, pull-ups, sit-ups, stretching exercises, and conditioning drill.

## WHAT TO TAKE AND WHAT NOT TO TAKE TO BOOT CAMP

It is important that you have certain documents when you arrive at boot camp: your Social Security card and, if married, your marriage license and the birth certificate of any children you have. Don't go to boot camp wearing a suit and tie; just wear neat, casual, comfortable clothes. Your drill instructor will see that you get everything else you need during training. Women should wear slacks or jeans, a blouse, and a sweater or jacket depending on the season. Uniform items will be issued within the first two days of your arrival; your civilian clothes will be locked in your luggage and stored for the duration of training. You should not bring much cash with you; on arrival you will receive coupons to purchase items in the Marine Corps Exchange.

Your storage space in the training barracks will be limited to room for your uniforms and basic necessities. Therefore, you should travel light when reporting to boot camp. The following items should not be taken to basic training; they will be confiscated upon arrival.

Alcoholic beverages or products, including any liquids containing alochol.

Food products, tobacco products (including cigarettes), chewing gum.

Firearms, ammunition, explosives, fireworks.

Lethal weapons, including blackjacks, brass knuckles, knives, straight razors, scissors.

Narcotics or associated equipment.

Commercial medication, including aspirin, ointments, laxatives, vitamins. Should you require medication, it will be given to you at boot camp.

Toiletries. These will be provided.

Products contained in glass.

Pressurized containers.

Sunglasses, unless prescribed by a doctor.

Playing cards, dice, or other gambling equipment.

Obscene or subversive literature or pictures.

Books, magazines, newspapers (except religious scriptures).

Cameras, radios, watches, jewelry (except wedding bands and religious medallions).

If you have a question about taking specific items with you, check with your recruiter.

### PROCESSING/FORMING

You will arrive at the recruit depot on a bus and be taken to a Receiving Barracks. There you will be met by a drill instructor, who will immediately start your processing into the Recruit Training Regiment. From that moment on, you will always be on the go; every hour is scheduled and programmed. The processing/forming period lasts about seven days for men recruits and about nine days for women recruits. Processing procedures are similar for men and women with minor differences. The Marine Corps understands that this is a critical time in the transition from civilian to military life. It is a critical time for you; you want to ensure that you get off to a good start. Do what you are told to do, and do it quickly to the best of your ability. Some of the things you will do in this period include the following:

Administrative Process
Hygienics
   Haircuts
   Showers

Basic Uniform Issue and Toilet Articles
Physical Screening
   Hearing Test
   Inoculations
   Eye Examinations
   Blood Test
   Urinalysis (positive results on pregnancy or drugs bring immediate discharge)
   Classification (ASVAB)
   Additional Administrative Processing
      Initiate Service Record Books
      Serviceman's Group Life Insurance
      G.I. Bill
   Training Film
   Introduction to Chain of Command
   Eye Examination
   Dental Processing
   Military Law
   Customs and Courtesies

## TRAINING

After processing is completed, training begins in earnest. The training is broken down into three phases. Phase I covers the 2nd to the 4th week; Phase II covers the 5th to the 7th week; and Phase III covers the 8th to the 11th week. Most of the subjects presented in these phases are listed below. The subjects are the men's subjects; women's subjects vary slightly. Women are not allowed to enter combat positions and do not receive instruction in offensive tactics.

*First Training Phase*
Initial Strength Test
Physical Training
   Log Drills
   Table Pt (Calisthenics)
   Obstacle Course
   Circuit Course
   Confidence Course
Bayonet Training/Close Combat

Instruction
    Bayonet Assault Course
    Pugil Stick Fight
Academics
    Marine Corps History
    Leadership Instruction
    First Aid
    M-16A2 Rifle
    Code of Conduct
    Interior Guard
    Uniform Regulations
    Troop Information
Swim Qualifications

Dental Hygiene
Senior Drill Instructor's Inspection
Initial Drill Evaluation
Physical Fitness Test
First Phase Academic Testing

*Second Training Phase*
Rifle Qualification
    Snapping-In
    Firing
Physical Training
    Stretching Exercise
    4-Mile Run
Drill
Mess and Maintenance

*Third Training Phase*
Dental Hygiene
Clothing Appointments
    Issue Service Uniforms
Physical training
    Table Pt
    Obstacle Course
    Log Drills
    5-Mile Motivation Run
    Circuit Course

Confidence Course

Close Combat
   Pugil Stick Training
   Hand-to-Hand Combat

Swim Qualifications

Nuclear, Biological, and Chemical Defense

Academics

Uniform Regulations

Information Classes

Individual Combat Training
   Grenades
   Infiltration Course
   Rappelling
   Squad Tactics

Practical Examination

Transition Training

Morning Colors Ceremony (Parris Island only). During Phase 3 a platoon will fold the holiday flag for Sunday night colors.

Series Commander's Inspection

Friday Parade

Company Commander's Inspection

Physical Fitness Test

Final Drill Evaluation

Battalion Commander's Inspection

5-Mile Endurance Run

Special Pay

Field Meet

Recruit Liberty

Final Pay

The Marine Corps in 1988 undertook extensive changes in its basic enlisted recruit training. With the goal of ensuring that every recruit, male and female, is equipped with basic combat skills, the Corps' new training syllabus incorporates Basic Warrior Training. This specialized training provides the Marine the skills required to function in a combat environment and to contribute more effectively to combat operations.

*Marine sergeants are interested in the success and welfare of their recruits.*

The program was initiated at the basic level by restructuring boot camp without affecting the length of training. Essentially, it provides more field training with more of a combat orientation. Recruits receive more training in organic infantry weapons, communications, land navigation, field firing, security measures, individual tactical measures, and the threat. To accommodate the new training, some recruit topics were reduced in hours.

Plans call for expanding Basic Warrior Training within the next few years to include improved training of infantrymen, additional training at the School of Infantry for non-infantrymen, improved small unit leader training with an emphasis on combat skills, and increased emphasis on combat preparedness in unit training to sustain individual combat skills.

The restructuring was initiated at the direction of the Commandant of the Marine Corps, in view of the changing threat situations that will place more Marines in increased jeopardy in the event of a conflict.

## MARINE TERMINOLOGY

The Marine Corps, like the other services, has its own terminology. Below are some of the terms you will encounter in boot camp.

| Marine Term | Meaning |
|---|---|
| Ashore | Off station. Where you go on leave or liberty |
| Aye Aye, Sir | Official acknowledgment of an order |
| Barracks | Building where Marines live |
| Below | Downstairs |
| Bivouac | An area where you pitch tents in the field to stay overnight |
| Blouse | Coat |
| Boondocks | Woods or wilds; training area |
| Brightwork | Brass or shiny metal; i.e., water faucets, doorknobs, etc. |
| Bulkhead | Wall |
| Bunk or rack | Bed |
| Chit | A small piece of paper; a receipt or authorization |
| CMC | Commandant of the Marine Corps |
| CO | Commander, Commanding Officer |
| Colors | A national flag |
| Cover | Hat |
| Cruise or tour | Period of enlistment |
| Deck | Floor |
| Drill | March |
| Esprit de corps | Spirit of camaraderie |
| Field | Boondocks where you train |
| Field day | Clean up an area |

| | |
|---|---|
| Galley | Kitchen |
| Gangway | Move out of the way or make room |
| Gear Locker | Storage room or locker for cleaning purposes |
| Gung ho | Working together; in the spirit |
| Hatch | Door |
| Head | Bathroom |
| Ladder | Stairs |
| Leave | Authorized vacation |
| Liberty | Authorized free time, but not leave |
| MOS | Military occupational specialty |
| NCO | Noncommissioned officer |
| NCOIC | Noncommissioned officer in charge |
| Overhead | Ceiling |
| Passageway | Corridor or hallway |
| Porthole | Window |
| PFT | Physical fitness test |
| PX | Post Exchange; comparable to a civilian department store |
| Quarters | A place to live; i.e., house, barracks, etc. |
| Reveille | Time to get up |
| Secure | Stop work, put away, close or lock |
| Scuttlebutt | Water fountain, rumors |
| Snapping in | Practicing getting into firing position |
| Squadbay | Large room in barracks where Marines live |
| Square away | Straighten up, make neat |
| Survey | Turn in unserviceable items |
| Swab | Mop |
| Taps | Time to sleep |
| Topside | Upstairs |
| W.M. | Woman Marine |

## MAIL/MAILING ADDRESS

Mail call is held once a day except on Sundays. You will be given time to write letters. Mail call is something you will look forward to while in training. Encourage your friends, parents, and relatives to write you at boot camp. Your mailing address while in boot camp will be:

Pvt (your name)
Plt (your platoon), company (your company designation)

_____ Battalion, Recruit Training Regiment
Marine Corps Recruit Depot
San Diego, California 92140
         OR
Parris Island, South Carolina 29905

## PHYSICAL CONDITIONING PRIOR TO BOOT CAMP

The most important thing to take to boot camp is a positive attitude. Boot camp is tough. But each year thousands of young men and women like you successfully complete it. Keep that in mind. One of the greatest challenges is the physical conditioning program. It is designed to build a strong body to go along with your positive attitude...an unbeatable combination.

To prepare yourself, start getting in shape as soon as you can. Begin gradually by running a mile a day and doing plenty of pull-ups and sit-ups. A Marine Corps pull-up starts with the palms either toward or away from the body and the body hanging straight down; it is completed when the chin is raised above the bar and the body hangs straight down. A sit-up starts with the feet flat on the mat, the legs bent and the hands behind the head. The feet are held by a partner. The sit-up is completed when the body is raised to bring the forehead, with hands still behind the head, directly

**Physical Fitness Chart**

| Points | Pull-ups | Sit-ups | 3-mile run | Points | Pull-ups | Sit-ups | 3-mile run | Points | Pull-ups | Sit-ups | 3-mile run |
|---|---|---|---|---|---|---|---|---|---|---|---|
| 100 | 20 | 80 | 18:00 | 66 | | 63 | 23:40 | 32 | | 32 | 29:20 |
| 99 | | | 18:10 | 65 | 13 | | 23:50 | 31 | | 31 | 29:30 |
| 98 | | 79 | 18:20 | 64 | | 62 | 24:00 | 30 | 6 | 30 | 29:40 |
| 97 | | | 18:30 | 63 | | | 24:10 | 29 | | 29 | 29:50 |
| 96 | | 78 | 18:40 | 62 | | 61 | 24:20 | 28 | | 28 | 30:00 |
| 95 | 19 | | 18:50 | 61 | | | 24:30 | 27 | | 27 | 30:10 |
| 94 | | 77 | 19:00 | 60 | 12 | 60 | 24:40 | 26 | | 26 | 30:20 |
| 93 | | | 19:10 | 59 | | 59 | 24:50 | 25 | 5 | 25 | 30:30 |

| | | | | | | | | | |
|---|---|---|---|---|---|---|---|---|---|
| 92 | | 76 | 19:20 58 | | 58 | 25:00 24 | | 24 | 30:40 |
| 91 | | | 19:30 57 | | 57 | 25:10 23 | | 23 | 30:50 |
| 90 | 18 | 75 | 19:40 56 | | 56 | 25:20 22 | | 22 | 31:00 |
| 89 | | | 19:50 55 | 11 | 55 | 25:30 21 | | 21 | 31:10 |
| 88 | | 74 | 20:00 54 | | 54 | 25:40 20 | 4 | 20 | 31:20 |
| 87 | | | 20:10 53 | | 53 | 25:50 19 | | 19 | 31:30 |
| 86 | | 73 | 20:20 52 | | 52 | 26:00 18 | | 18 | 31:40 |
| 85 | 17 | | 20:30 51 | | 51 | 26:10 17 | | 17 | 31:50 |
| 84 | | 72 | 20:40 50 | 10 | 50 | 26:20 16 | | 16 | 32:00 |
| 83 | | | 20:50 49 | | 49 | 26:30 15 | 3 | 15 | 32:10 |
| 82 | | 71 | 21:00 48 | | 48 | 26:40 14 | | 14 | 32:20 |
| 81 | | | 21:10 47 | | 47 | 26:50 13 | | 13 | 32:30 |
| 80 | 16 | 70 | 21:20 46 | | 46 | 27:00 12 | | 12 | 32:40 |
| 79 | | | 21:30 45 | 9 | 45 | 27:10 11 | | 11 | 32:50 |
| 78 | | 69 | 21:40 44 | | 44 | 27:20 10 | 2 | 10 | 33:00 |
| 77 | | | 21:50 43 | | 43 | 27:30 9 | | 9 | 33:10 |
| 76 | | 68 | 22:00 42 | | 42 | 27:40 8 | | 8 | 33:20 |
| 75 | 15 | | 22:10 41 | | 41 | 27:50 7 | | 7 | 33:30 |
| 74 | | 67 | 22:20 40 | 8 | 40 | 28:00 6 | | 6 | 33:40 |
| 73 | | | 22:30 39 | | 39 | 28:10 5 | 1 | 5 | 33:50 |
| 72 | | 66 | 22:40 38 | | 38 | 28:20 4 | | 4 | 34:00 |
| 71 | | | 22:50 37 | | 37 | 28:30 3 | | 3 | 34:30 |
| 70 | 14 | 65 | 23:00 36 | | 36 | 28:40 2 | | 2 | 35:00 |
| 69 | | | 23:10 35 | 7 | 35 | 28:50 1 | | 1 | 36:00 |
| 68 | | 64 | 23:20 34 | | 34 | 29:00 | | | |
| 67 | | | 23:30 33 | | 33 | 29:10 | | | |

*A proud Marine private first class is congratulated by a sergeant.*

above or in front of the knees and then returned to the mat. Increase the distance and speed you run and the number of each
exercise each week. Work out regularly and be consistent. The
Physical Fitness Chart is used to determine your score in and out
of recruit training. Check your progress against it. It will give you a
good idea of how well you will do once you get to boot camp.

*Marine training is strenuous, but it creates "a few good men"—and women.*

The chart assigns a point value to each of the three events. The maximum obtainable score for any one event is 100 points, and 300 points represents a perfect score.

Physical conditioning is also emphasized at women's boot camp. Several training hours are devoted to physical fitness. The Physical Fitness Test (PFT) consists of bent-knee sit-ups, a flex-arm hang, and a 1½-mile timed run. To pass the PFT you must run the 1½ miles in less than 15 minutes, hang from the bar for a minimum of 16 seconds, and complete the 22 sit-ups in 1 minute. It is to your advantage to prepare yourself physically for recruit training. Both men and women must meet height and weight standards to be allowed to stay in the Marine Corps.

## CODE OF CONDUCT

The Code of Conduct is a creed by which all military members must serve. It applies to all services at all times, in both peace and war. You will probably have to memorize the code in basic training. It consists of six articles:

Article I

I am an American fighting man. I serve in the forces which guard my country and our way of life. I am prepared to give my life in their defense.

Article II

I will never surrender of my own free will. If in command I will never surrender my men while they still have the means to resist.

Article III

If I am captured, I will continue to resist by all means available. I will make every effort to escape and aid others to escape. I will accept neither parole nor special favors from the enemy.

Article IV

If I become a prisoner of war, I will keep faith with my fellow prisoners. I will give no information or take part in any action which might be harmful to my comrades. If I am senior, I will take command. If not, I will obey the lawful orders of those appointed over me and will back them up in every way.

Article V

When questioned, should I become a prisoner of war, I am required to give only name, rank, service number, and date of birth. I will evade answering further questions to the utmost of my ability. I will make no oral or written statements disloyal to my country and its allies or harmful to their cause.

Article VI

I will never forget that I am an American fighting man, responsible for my actions, and dedicated to the principles which make my country free. I will trust in my God and in the United States of America.

## *MEDICAL CARE*

Medical care and dental services are available 24 hours a day. Preventive inoculations are given throughout training as necessary. You will receive a medical screening early in training to ensure that you are able to perform satisfactorily from a physical standpoint. Sleep is an absolute right of all recruits. The sleeping period is an uninterrupted period of 8 hours, beginning with taps and ending with reveille. The only exceptions are those circumstances when a recruit is required to perform guard duty, fire/security watch, or during field training skills week.

## *PAY*

You will not need much money during boot camp, because you will not have the time or the opportunity to spend money. You will receive coupons upon arrival to buy needed items. You will receive additional coupons once more during the training, and then you will receive all your accumulated pay the week of graduation. You will receive over $600 a month before deductions.

You probably have heard or read advertisements that the Marines are looking for a few good men or a few good women. Attrition rates in Marine Corps basic training are higher than in the other services. Men and women who have gone through boot camp successfully comment on how tough it was and also how proud of themselves they are for completing the training. They develop pride, esprit, and confidence. The Marines are something special, and they live up to their motto of Semper Fidelis—Always Faithful; faithful to yourself, your comrades, your Corps, your country, and your God.

Chapter **VIII**

---

# The US Coast Guard

## MISSION

The Coast Guard performs its mission of protecting America's coastlines and inland waterways by enforcing customs and fishing laws, combatting drug smuggling, conducting search and rescue operations, maintaining lighthouses, and promoting boat safety. The Coast Guard is part of the Department of Transportation; in time of war it may be placed in the Department of Defense under the command of the Navy. A vital part of the armed services, the Coast Guard has participated in every major American military campaign. Its workforce of about 5,000 officers and over 32,000 enlisted personnel perform in many different occupations to support the mission of the Coast Guard. Each year the Coast Guard has openings for about 5,000 enlistees in a wide range of challenging careers. In recent years the Coast Guard has played an expanding role in environmental protection with the development of sophisticated techniques to prevent and detect oil spills. Its mission has also been expanded in maritime law enforcement in combatting drug traffic. A Coast Guard career is a career of action.

## ENLISTMENT AND INITIAL TRAINING

Applicants for enlistment in the Coast Guard must be in good health, possess good moral character, and make the minimum required scores on the Armed Services Vocational Aptitude Battery (ASVAB). Coast Guard regular enlistments are for four years of active duty. Provided class openings are available, qualified applicants can be guaranteed their choice of specific occupational training under the Coast Guard's Guaranteed School Program. Qualified applicants may also enlist up to 12 months before beginning active duty. Coast Guard recruits must be at least 17 years old and must not have reached their 26th birthday on the day of enlistment.

Two types of training are provided to Coast Guard recruits: recruit training and job training. After completing the enlistment process, all Coast Guard recruits attend recruit training or boot camp at Cape May, New Jersey. Boot camp lasts approximately eight weeks; it is designed to provide a transition from civilian life to that of service with the Coast Guard. The course is demanding, both physically and mentally. Coast Guard recruit training instills in each trainee a sense of teamwork and discipline. Coast Guard history, missions, customs, and basic discipline are all part of the training course. Boot camp includes physical training, classroom work, and practical application of the subjects studied.

The Coast Guard maintains basic petty officer (Class A) schools for formal training in specific occupational specialties. Courses of study in these Class A schools vary from 10 to 42 weeks depending on the rating or specialty area taught. Each school provides a course of study that leads to advancement to the third class petty officer level. Specialty schools in the other services can be used by Coast Guard personnel in addition to, or in place of, Coast Guard school for training in certain ratings. Upon successful completion of Class A school, the graduate becomes a qualified specialist and can expect assignment to a field unit for duty and additional on-the-job training in his or her specialty.

Opportunities for additional professional training are available to qualified career-oriented personnel in the form of advanced petty officer (Class B) and special (Class C) schools. These advanced schools range in duration from a few weeks to several months depending on the skills taught. Senior enlisted personnel in certain ratings are also eligible to compete for assignment to special degree programs within their occupational specialty areas.

## *WHAT TO EXPECT*

Just as in the other military services, Coast Guard recruit training is very demanding. Ths staff will push you to the limits of your endurance, and you will accomplish more than you previously thought possible. Although you will think about giving up, you must keep your sights set on graduation. You must continually put forth your best efforts. In eight weeks you will develop increased self-confidence and self-esteem and will become a contributing member of the US Coast Guard.

The stated mission of recruit training is as follows: "The mission of the Coast Guard Recruit Training Center is to train young men and women to be always ready to serve in the sky and at sea, with courage, alertness and devotion, eager to learn the skills and traditions of the United States Coast Guard in the service of their country and humanity." Service is the main feature of the Coast Guard.

## *INITIAL PROCESSING*

Recruit training takes eight weeks. Usually you report on a Monday morning, and the rest of the week is spent in initial issue of uniforms, medical and personnel forms, medical and dental examinations, first classes on Coast Guard rules and regulations, urinalysis (controlled substance abuse will result in discharge), piece (rifle) issue, inoculations, haircut, trip to the base exchange for needed personal items, physical training, and other administrative requirements. You are placed in companies under a company commander. Under the direction of the company commander and his or her assistants, the company is paced through all the training phases until graduation or transfer to the field. On the following Monday training begins in earnest as you start attending classes in all aspects of Coast Guard operations.

Training consists of instruction and practical experience in:

Customs and Courtesies
Military Conduct
Watch, Quarter, and Station Bill
Boat and Deck Seamanship
Marlinspike Seamanship
Military Drill
Uniforms
Career Information
First Aid and Personal Hygiene
Survival Safety
Military Orientation/Indoctrination
Coast Guard History
Damage Control
Small Arms
Vessel and Aircraft Characteristics
Surface Preservation
Signals

Watchstanding
Communications
Basic Engineering
Physical Fitness
Drug and Alcohol Abuse
Fiscal Responsibility
Sex Education
GI Bill/Education Benefits
Chapel/Crisis Intervention Orientation
Civil Rights

It is a full agenda with emphasis on seamanship.

### Earning a Greenbelt

Every recruit must earn a greenbelt by the start of the fourth week of training. To earn a greenbelt, you must recite verbatim some of the information given below. You must also show that you have mastered some basic skills. The following list indicates what you must accomplish. Your Company Commander (CC) will schedule a time for you to take your greenbelt test. You will not be allowed to proceed farther in your training until you pass it.

Since there is a lot of information to learn, you are encouraged to try to earn the greenbelt as soon as possible, preferably before the beginning of week three. You will be introduced to all this information as soon as you arrive at recruit training. As a head start six of these items are included in this chapter.

1. The Mission of Recruit Training
2. Nautical Terminology and Military Time
3. Eleven General Orders
4. Chain of Command
5. The Position of Attention
6. How and Where to Salute
7. Piece Number (Rifle Serial Number)
8. How to Answer a Telephone
9. What to Do in Case of Fire
10. Duties of the Quarterdeck/Security Watch
11. How to Enter and Leave a Compartment
12. Phonetic Alphabet
13. How to Make a Rack (Bed)
14. How to Set Up a Locker

15. Plan of the Day Information
16. Information Posted on the Squadbay Bulletin Board
17. The Manual of Arms

## NAUTICAL TERMINOLOGY

Here are some of the nautical terms you will have to learn. You will encounter many more while serving in the Coast Guard.

| Coast Guard Term | Meaning |
|---|---|
| Belay, or avast | Stop |
| Below | Downstairs |
| Bulkhead | Wall |
| Colors | American flag |
| Compartment | Room |
| Cover | Hat |
| Deck | Floor |
| Head | Bathroom |
| Ladder | Stairs |
| Overhead | Ceiling |
| Quarterdeck | Area designated by the CO for official and ceremonial functions |
| Rack | Bed |
| Scuttlebutt | Drinking fountain; rumor |
| Sir, aye aye Sir | I understand your order and will comply |
| Swab | Mop |
| Topside | Upstairs |

## MILITARY TIME

Time is told on a continuous 24-hour clock. Rather than distinguishing between morning and afternoon, the time is read sequentially from 0001 to 2400. For example, fifteen minutes past midnight is written 0015 and spoken as zero zero fifteen. Half past one in the morning is written 0130 and spoken as zero one thirty. Two o'clock in the afternoon is two hours after twelve and therefore is written 1400 and spoken as fourteen hundred. Quarter to ten in the evening is written 2145 and spoken as twenty-one forty-five.

*A Coast Guard review is a proud moment for officers and recruits alike.*

## ELEVEN GENERAL ORDERS

The eleven General Orders are common to all branches of the armed forces. The orders are modified for Coast Guard use.

1. To take charge of my watch and all government property in view.
2. To stand my watch in a military manner, keeping always on the alert and observing everything that takes place within sight or hearing.
3. To report all violations of orders I am instructed to enforce.
4. To report all alarms, calls for assistance, and safety hazards to the Officer of the Day.
5. To quit my watch only when properly relieved.
6. To receive, obey, and pass on to the watchstander who relieves me all orders from the Commanding Officer,

Officer of the Day, and officers and petty officers of the watch section.

7. To talk to no one except in the line of duty.
8. To give the alarm in case of fire or disorder.
9. To call the Duty Petty Officer in any case not covered by instructions.
10. To salute all officers and all colors and standards not cased.
11. To be especially watchful at night and to allow no one to pass without proper authority.

These general orders must be recited verbatim in the greenbelt examination.

## CHAIN OF COMMAND FOR RECRUITS

The chain of command is a structure of authority and responsibility that exists in all the services. Everyone in the Coast Guard is responsible for his or her actions to someone senior in either rank or position. The order of seniority due to position is called the chain of command. The chain of command is needed so that everyone knows how and where he or she fits into the organization relative to everyone else. Your immediate senior is your Company Commander (CC). He/she must answer to the Battalion Adjutant. The Battalion Adjutant's supervisor is the Military Training Officer. The full sequence is given below. You must always follow the chain of command when making a request. You must know the name and rank of each person in the chain of command.

Company Commander (CC) _____
Battalion Adjutant (BA) _____
Military Training Officer (MTO) _____
Regimental Officer (RO) _____
Training Officer (TO) _____
Executive Officer (XO) _____
Commanding Officer (CO) _____
Commandant of the Coast Guard _____
Secretary of Transportation _____
Commander in Chief (President of the US) _____

Every unit within the Coast Guard is subdivided into departments and divisions for organizational efficiency. Training Center Cape May has five major divisions: Training, Comptroller, Facili-

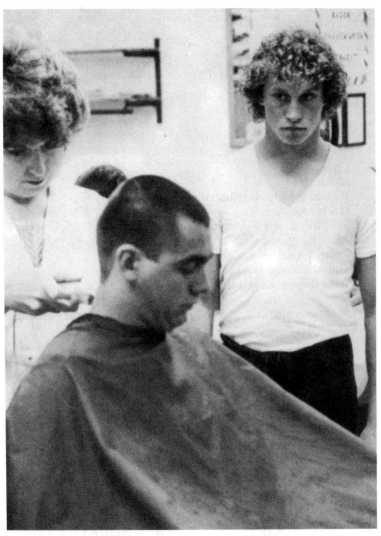

*Shaggy locks fall victim to the trim look as Coast Guard recruits prepare for training.*

ties Engineering, Administration, and Health Services. The recruit training program is a Training Division function. Training Division is further divided into five branches: Instruction and Testing, Seamanship, Armory, Military Training, and Physical Training. During your training, you will meet Coast Guardsmen from each of these branches.

## THE POSITION OF ATTENTION

Sir, the position of attention is:
Head erect,
Eyes in the boat,
Chin in,
Shoulders back,
Chest out,
Stomach in,
Weight evenly distributed on both feet,
Heels together,
Feet forming a 45-degree angle,
Arms hanging naturally at sides,
Palms facing inwards,
Thumbs along trouser seams,
With fingers joined in a natural curl, sir.

This must also be recited verbatim during the greenbelt examination.

## PHONETIC ALPHABET

| | | | |
|---|---|---|---|
| A | ALFA | N | NOVEMBER |
| B | BRAVO | O | OSCAR |
| C | CHARLIE | P | PAPA |
| D | DELTA | Q | QUEBEC |
| E | ECHO | R | ROMEO |
| F | FOXTROT | S | SIERRA |
| G | GOLF | T | TANGO |
| H | HOTEL | U | UNIFORM |
| I | INDIA | V | VICTOR |
| J | JULIET | W | WHISKEY |
| K | KILO | X | XRAY |
| L | LIMA | Y | YANKEE |
| M | MIKE | Z | ZULU |

This phonetic alphabet is used by all military services.

## RESTRICTIONS ON RECRUITS

When in groups of four or more, recruits must march in formation. When in groups of three or less, recruits must double time, except when they are on light duty, wearing a dress uniform, during Divine Hours (Sundays), or fifteen minutes before and after any meal. Recruits may not go into any squadbay other than their own unless they are required to do so in the course of their watchstanding duties. If you are a smoker now, try to give up tobacco before you go to boot camp. Smoking is not allowed in recruit training. It is not a good idea to wait until you arrive to try to give up smoking. The training is too demanding to take on the additional burden of stopping smoking "cold turkey."

## CONDUCT WHILE ON LIBERTY

Liberty is a privilege that is earned and not automatically given to everyone. While on liberty you represent the US Coast Guard and should show pride in your service. The use of illicit narcotic substances is absolutely prohibited. Anyone caught using drugs, or found to have used drugs, will be court-martialed and discharged from the Coast Guard. The consumption of alcohol by anyone under 21 years of age is not permitted. Excessive consumption of alcohol and underage drinking will be punished according to current command policy. Uniforms must be worn properly and with respect. Your conduct must reflect pride in the Coast Guard. Horseplay, swearing, and spitting are not permitted and will be dealt with severely.

## AUTHORIZED NON-ISSUE ITEMS

The following non-issue items are authorized for recruits:

All-white athletic shoes
Stationery
Pens/pencils
Wallet
Address book
Stamps
Iron
Religious materials
Long underwear (tops/bottoms) white only (Jan–Apr)

Personal Hygiene items:
Toothpaste
Toothbrush
Soap
Shampoo*
Cartridge type razor
Shaving cream
Manicure kit
Comb/brush
Dental floss
Medication prescribed by Training Center dispensary

Men only:
Electric razor
Jockey-type shorts (white only)

Women only:
Hair dryer
Curlers
Pajamas
Bathrobe
Makeup*
Feminine hygiene items
Underpants (white or flesh-color)
Bra (white or flesh-color; at least 2 sports bras recommended)

*Glass containers of any type are prohibited unless issued by dispensary for medicine.

Any nonauthorized items found will be confiscated and disposed of. If you report to the center with any unauthorized items, you may request that they be held until graduation. Any recruit found concealing contraband items is subject to disciplinary action.

## MAIL AND TELEPHONE

Mail is delivered to the Training Center Monday through Friday. Due to the volume of mail, it may take an extra day to reach you. Company mail orderlies pick up mail at a time set by the Company Commander (CC) and return it to the CC for inspection. Any suspicious looking letters or packages are opened in the presence of the CC. Contraband items are seized. Food items of any kind, magazines, newspapers, books, and drugs of any kind are prohibited. Outgoing mail may be sent from the drop box at the Post Office or dropped in the mailbox located between Healy Hall and the gym. Mail is picked up twice a day. Your mailing address while at the Training Center is:

SR _____
Recruit Company _____
Healy Hall, OR Munro Hall, OR James Hall
USCG Training Center
Cape May, NJ 08204-5002

There are telephones in the barracks for recruit use. The telephones are either coin or credit card/collect pay telephones. Phone privileges are granted by the CC to recruits who perform satisfactorily.

## FRATERNIZATION POLICY

During recruit training many things are forbidden. One item of particular importance is the prohibition of fraternization between recruits and permanent party personnel, and between male and female recruits. Fraternization is defined as any association between individuals that is not authorized by the regulations or in the course of duty. Recruits are required to avoid situations that would violate this policy. Recruits should not engage in idle conversation, accept favors from, or agree to do things on liberty, with any staff member. Recruits may not pursue romantic interests with other recruits or staff members. Recruits are required to report any incident involving fraternization. Failure to report an incident is a punishable offense.

## RELIGIOUS SERVICES AND COUNSELING

Many recruits find religious services comforting and a good way to ease the pressures of training. The Training Center chapel is staffed by chaplains of various faiths and holds services every Sunday morning which you may attend. You may also join the recruit choir. Recruits from each company are assigned as company religious representatives and attend weekly meetings, lead company prayers, and work in the chapel at various times.

The training program is at times very strenuous. This, combined

*The counsel of a chaplain is always available to Coast Guard recruits.*

with the stress of separation from family and friends, often creates a need for recruits to seek counseling. Seeking counseling is not a sign of weakness. In many instances counseling helps immensely with problems. There are several people you can see to discuss your problems. The first and most obvious is your company commander, who has had a lot of experience in dealing with recruits. If you feel uncomfortable discussing personal problems with your CC, you can make an appointment to see a chaplain. Your Company Commander will arrange an appointment without asking you any questions. Other counseling resources include a staff psychiatrist and a human relations counselor. Women may wish to speak with a female social worker or nurse. Appointments to see these professional people are made through your Company Commander.

## PAY AND FINANCES

Although you accrue pay for the entire time you are in recruit training, you are not paid all your money until you depart. Periodically you are given a cash allowance for the purchase of incidentals. Just before graduation you receive your remaining pay.

This assures you that you have enough money to pay for travel and other expenses en route to your first unit. Recruits with dependents receive additional allowances at regular intervals so that they can send money home. Your CC will make the necessary arrangements with the pay office. Money is deducted from your pay to cover the costs of laundry, ditty bags (necessary items that are not provided by the Coast Guard), and haircuts. You can refer to the pay chart in Chapter III to get an idea of your pay.

## EXCHANGE

You may use the base exchange when authorized by your CC. The exchange stocks most items needed on a day-to-day basis. Additionally, the exchange will hold items on layaway until your graduation day. During training you may not buy any food, magazines, or other contraband items. You may buy toilet articles, irons, towels, uniform clothing, and miscellaneous items such as stationery, notebooks, paper, and pens.

## GRADUATION WEEK PREPARATION

Preparation for graduation begins on Wednesday of your eighth week of training. The final three days in training are hectic; there are many things you must do before you leave Cape May. First you are asked to prepare a critique of the training program for the Commanding Officer. The CO is looking for useful, constructive criticism. If you had any problems, good or bad experiences, or if you want to inform the CO about any particular person, this is the opportunity to do it. Next you need to make travel arrangements for your trip home on leave and to your first duty station. Travel arrangements may be made anytime before the morning you graduate. On graduation day you pick up and pay for your tickets. If you want to make any changes, they can be made after graduation.

Visitors are allowed to come aboard the Training Center commencing at 0900 on graduation day. A short movie presentation is given in the auditorium at 1000, which briefly describes the training program. Graduation day will be one of the proudest moments in your life. You will have developed lasting friendships. You will have learned to depend on your classmates. You will be an important part of a specialized team where everyone relies on everyone else; where everyone helps everyone else. It is the Coast Guard way.

# Conclusion

You will find useful information regarding military occupations in the Appendix. Basic training or recruit training is only your first step. Basic training is a difficult time, but it is relatively short. You must also look ahead and plan what you want to do in the rest of your military career, whether it is just to fulfill your obligation or to serve a full 30 years. You will find lists of military occupations in each service as well as how those military occupations equate with civilian occupations. Many young people are not sure what they want to do with their lives when they graduate from high school. In fact, many people are not sure what they want to do with their lives when they graduate from college. Many feel that they need discipline and a sense of purpose in their lives, and the military provides that.

It is important to recognize the growing importance of the contributions women are making to national defense. Less than fifteen years ago women made up only 2 percent of active duty personnel; today women make up about 10 percent of active duty personnel. The scope of women's opportunities in the military has also expanded. Women are currently eligible to enter almost 90 percent of all military job specialties. Examples of the many occupations women are now entering include helicopter mechanic, missile maintenance technician, and heavy equipment operator. According to federal laws and policies, women may not be assigned to duty that involves a high exposure to direct combat. Women would not be assigned to such duties as tank crew member, fighter pilot, submarine crew, and infantryman. However, despite these restrictions, the commitment to integrate women into the military has never been higher. The outlook for women officers and enlistees in the military suggests that the future will provide even greater opportunities.

Probably the single most important thing you need to carry you through basic training successfully is a positive attitude that you are ready for basic and that you will get through no matter how tough it sometimes gets. You will have a positive attitude if you know what you are going to encounter at basic and if you have worked hard to get yourself in good physical condition. You are not expected to know everything that will be presented in recruit training; you are expected to keep an open mind and work hard at learning. All of basic training is a learning experience and it should be accepted that way.

Go into the military with your eyes open. Find out as much as you can about the particular service before you make any commitment. Look farther ahead than basic training. Plan on obtaining as much additional education as you can while in the military. Set your sights high. Look into the advantages of serving your country as an officer. You are embarking on a great adventure. Make the most of it.

# Appendix

## Army Occupations

Career fields are listed alphabetically, beginning with those combat-related groups closed to women (indicated by "C"). Those marked with an asterisk (*) are generally open to all applicants, but may include specific combat-related positions available only to men.

| Career Fields | Duties & Responsibilities | Qualifications | Examples of Civilian Jobs |
|---|---|---|---|
| Armor (C) | Performs combat tasks using tanks and armored reconnaissance vehicles. | Team sports, mechanical maintenance, orienteering. | Heavy equipment operator supervisor. |
| Combat Engineering (C) | Constructs and maintains roads and bridges; operates powered bridges; constructs minefields and installs booby traps, demolitions with high explosives; erects temporary shelters and sets up camouflage. | Automotive mechanics, carpentry, woodworking, mechanical drawing and drafting courses. | Blaster, construction equipment operator, construction supervisor, bridge repairer and lumber worker. |
| Infantry (C) | Performs combat tasks using rifles, mortars, tank destroying missiles, personnel carriers, vehicle-mounted guns and fire control equipment. | Team sports, orienteering, hunting and other outdoor sports. | Supervisor, gunsmith, security officer, firearms handler. |
| Administration | Performs general administrative duties such as typing, stenography and postal functions and specialized administrative duties | Basic clerical and communication abilities, typing, bookkeeping, stenography or office management skills desirable. | Clerk typist, secretary, employment interviewer, postal clerk, recreation specialist, office manager, personnel clerk, |

| | | | |
|---|---|---|---|
| | such as personnel, legal, club management, equal opportunity and chapel activities. | | bookkeeper, cashier, payroll clerk, court clerk, and restaurant or cafeteria manager. |
| **Aircraft Maintenance** | Performs the mechanical functions of maintenance, repair and modification of helicopters, turboprop, and reciprocating engine aircraft. | Considerable mechanical or electrical aptitude and manual dexterity. Shop mathematics and physics desirable. | Aircraft mechanic, plane inspector. |
| **Aircraft System Maintenance** | Performs maintenance of aircraft accessory systems, propulsion systems, armament systems, fabrication of metal materials used in aircraft structural repair and inspection and preservation. | Electrical and mechanical aptitude shop mathematics and shop work is desirable. | Aircraft mechanic, aircraft electrician, sheet metal machinist. |
| **Air Defense Artillery (*)** | Emplaces, assembles, tests, maintains and fires air defense weapons systems; operates fire control equipment, radars, computers, automatic data transmission and associated power supply equipment. | Basic mechanical, electrical, electronic, mathematical abilities; emotional stability and high degree of reasoning ability. | Map and topographical drafter, cartographer. |
| **Air Defense Missile Maintenance** | Inspects, tests, maintains and repairs guided missile fire control equipment and related radar installations which guide missile to target. | Mathematics, physics, electricity and electronics. | Radio installation and repair inspector, electronic equipment technician, radio and TV repairer. |
| **Ammunition** | Handles, stores, reconditions and salvages ammunition, explosives, and components; locates, removes | Mechanical aptitude, attentiveness, good close vision, normal color discrimination, | Toxic chemical handler, ammunition inspector and acid plant operator. |

| Career Fields | Duties & Responsibilities | Qualifications | Examples of Civilian Jobs |
|---|---|---|---|
| | and destroys or salvages unexploded bombs and missiles. | manual dexterity and hand-eye coordination. | |
| Audiovisual | Operation of radio and television equipment, still and motion picture photography, audiovisual equipment repair and graphic illustration. | Install, inspect and maintain radio and television equipment, prepare illustrations, film and slide processing. | Cameraman, camera repair, photography, illustrator and studio accessories. |
| Automatic Data Processing | Operates a variety of electric accounting and automatic data processing equipment to produce personnel, supply, fiscal, medical, intelligence and other reports. | Reasoning and verbal ability, clerical aptitude, finger and manual dexterity and hand-eye coordination. Knowledge of typing and office machines. | Coding clerk, key punch, computer and sorting machine operator, machines records unit supervisor. |
| Aviation Communications-Electronic Systems Maintenance | Repairs and maintains navigation, flight control, and associated communications equipment. | Electrical/electronic theory and repair. | Electronics technician, radar repairer, electrical instrument mechanic repairer. |
| Aviation Operations | Maintains, installs and repairs aviation communication radar systems used for aircraft navigation and landing; performs air traffic control duties. | Mathematics and shop courses in electricity and electronics useful. | Radio and television or electrical instrument repairer, communication, electrical and electronics engineer and radio engineer. |
| Ballistic/Land Combat Missile and Light Air Defense Weapons System Maintenance | Inspects, tests, maintains and repairs tactical missile systems and related test equipment and trainers. | Mathematics, physics, electricity, electronics (radio and TV) and blueprint reading. | Electronic equipment technician, radio electrician and mechanic, TV repair and service technician. |
| Band | Plays brass, woodwind or percussion instrument in | Instrumental audition on brass, woodwind or percussion | Bandsperson, bandmaster, musician, accompanist, arranger, |

| Category | Description | Courses helpful | Civilian counterparts |
|---|---|---|---|
| | marching, concert, dance, stage and show bands, combos or instrumental ensembles. Sings in vocal group, writes and arranges music. | instrument. | music director, orchestrator, music teacher, orchestra leader and vocalist. |
| Chemical | Provides decontamination service after chemical, biological or radiological attacks, produces smoke for battlefield concealment, repairs chemical equipment and assists in overall planning of chemical, biological, or radiological activities. | Biology, chemistry and electricity. | Laboratory assistant (biological, chemical or radiological), pumper and repairer (chemical) and exterminator. |
| Communications-Electronics Maintenance | Installs, maintains radar and radio receiving, transmitting carrier and terminal equipment. | Electricity, mathematics, electronics and blueprint reading. | Radio control room technician, radio mechanic, transmitter, radio and TV repairer. |
| Communications-Electronics Operations | Installs, maintains field telephone switchboards and field radio communications equipment. | Mathematics, physics and shop courses in electricity. | Communications engineer assistant, plant electrician and radio electrican or operator. |
| Electronic Warfare Cryptologic Operations | Collects and analyzes electromagnetic emissions; performs electronic warfare duties in fixed or mobile operations. | Verbal and reasoning ability and perceptual speed; aural and visual acuity. | Radio and telegraph operator, navigator, intelligence research analyst, statistician, signal collection technician. |
| Electronic Warfare Intercept Systems Maintenance | Installs, operates and maintains intercept, electronic measuring and testing equipment. | Physics, mathematics, electricity, electronics (radio and TV) and blueprint reading. | Electrical instrument repairer and electronic equipment inspector. |
| Finance and Accounting | Maintains pay records of military personnel; prepares vouchers for payment; prepares reports, | Dexterity in the operation of business machines. Typing, mathematics, statistics and basic | Paymaster, cashier statistical or audit clerk, accountant, budget clerk and bookkeeper. |

| Career Fields | Duties & Responsibilities | Qualifications | Examples of Civilian Jobs |
|---|---|---|---|
| | disburses funds; accounts for funds, to include budgeting, allocation, auditing; compiles and analyzes statistical data and prepares cost analysis records. | principles of accounting desirable. High administrative aptitude mandatory. | |
| Field Artillery (*) | Operates, maintains and directs fire of field artillery guns, howitzers, missiles, rockets and related weapons. Operates and maintains supporting equipment such as target acquisition radars, sound and flash ranging, meteorological and survey equipment. | Emotional stability, mathematics and reasoning abilities. | Map and topographical drafter, cartographer, surveyor, weather chart preparer. |
| Food Service | Plans regular and special diet menus, cooks and bakes food in dining facilities and during field exercises. Serves as aide and cook on personal staff of general officer. | Home economics, work in a restaurant, bake shop or meat market. | Cook, chef, caterer, baker, butcher, kitchen supervisor and cafeteria manager. |
| General Engineering | Provides utilities and engineering services such as electric power production, building and roadway construction and maintenance, salvage activities, airstrip construction, firefighting and crash rescue operations. | Mechanical aptitude, emotional stability and ability to visualize spatial relationships. Carpentry, woodworking or mechanical drawing. | Carpenter, construction equipment operator, electrician, firefighter, driver, plumber, welder, bricklayer. |
| Mechanical Maintenance | Services and repairs land and amphibious wheel and track | Automotive mechanics, electricity, blueprint reading, | Automotive mechanic, motor analyst, bakery or refrigeration |

|  | | | |
|---|---|---|---|
|  | vehicles ranging from cars and light trucks to heavy tanks and self-propelled weapons; installs and repairs refrigeration, bakery and laundry equipment. | machine shop and physics. | equipment, repairer, frame, wheel alignment and tractor mechanic. |
| **Medical** | Assists or supports physicians, surgeons, nurses, dentists, psychologists, social workers, and veterinarians in 32 separate job classifications. Some provide direct patient care in hospitals and clinics, others make and repair eyeglasses, dentures, or orthomedical equipment, or maintain medical records. All play significant roles in a modern worldwide health care delivery system. | Biology, chemistry, hygiene, sociology, general math, algebra, animal care; knowledge of mechanics and electronics; general clinical skills. | Social worker (case aide), practical nurse, nurse's aid, dental assistant, surgeon's assistant, psychological aide, hospital attendant or orderly, veterinary assistant, food quality control, medical equipment repairer, medical or dental laboratory technician, physical therapy assistant, dietetic technician. |
| **Military Intelligence** | Gathers, translates, correlates and interprets information, including imagery, associated with military plans and operations. | English composition, typing, foreign languages, mathematics and geography. | Investigator, interpreter, records analyst, research worker and intelligence analyst (government). |
| **Military Police** | Enforces military regulations; protects facilities, roads and designated sensitive areas and personnel; controls traffic movement; guards military prisoners and enemy prisoners of war; investigates traffic accidents and crimes involving military personnel. | Sociology and demonstrated prowess and leadership in athletics and other group work helpful. | Police officer, plant guard, detective, investigator, crime detection laboratory assistant and ballistics expert. |

| Career Fields | Duties & Responsibilities | Qualifications | Examples of Civilian Jobs |
|---|---|---|---|
| **Petroleum** | Receives, stores, preserves and distributes bulk packaged petroleum products; performs standard physical and chemical tests of petroleum products; storage and distribution of purified water. | Hygiene, biology, physics, chemistry and mathematics. | Biological laboratory assistant, petroleum tester, chemical laboratory assistant. |
| **Public Affairs** | Prepares and disseminates news releases on military activities; prepares scripts, newsletters, announcements and public speaking engagements. | Clerical aptitude, verbal ability, clear speech and attentiveness. | Newspaper editor, editorial assistant, public information center. |
| **Supply and Service** | Receives, stores and issues individual, organizational and expendable supplies, equipment and spare parts; establishes, posts and maintains stock records; repairs and alters textile, canvas and leather supplies, rigs parachutes, decontaminates materials. Performs mortuary and grave registration functions. | Mathematical ability and perceptual speed in scanning and checking supply documents. Verbal ability; courses in bookkeeping, typing and office machine operations. | Inventory clerk, stock control clerk or supervisor, shipping or parts clerk, warehouse manager, parachute rigger and funeral attendant. |
| **Topographic Engineering** | Performs land survey; produces construction drawings and plans, maps, charts, diagrams and illustrated material; constructs scale models of terrain and structures. Operates offset duplicators, presses and bindery | Mechanical drawing and drafting, blueprint reading, commercial art, fine arts, geography and mathematics. | Drafting (structural, mechanical and topographical), cartographic and art layout, model marker, commercial artist and physical geographer. |

| Career Fields | Duties & Responsibilities | Qualifications | Examples of Civilian Jobs |
|---|---|---|---|
| Transportation | equipment. Repairs survey instruments and reproduction equipment. Operates and performs preventive maintenance on passenger, light, medium and heavy cargo vehicles; operates and maintains marine harbor craft; performs as air traffic controller. | Mechanical aptitude, manual dexterity, eye/hand coordination, FAA certification for air traffic control, license for vehicle operation. | Truck driver, FAA air traffic controller. |

## Navy Occupations

Career fields are listed alphabetically, beginning with those combat-related groups closed to women (indicated by "C").

| Career Fields | Duties & Responsibilities | Qualifications | Examples of Civilian Jobs |
|---|---|---|---|
| Aviation Anti-submarine Warfare Operator (C) | Performs general flight crew duties; operates ASW sensor systems; performs diagnostic function to effect fault, isolation and optimize system performance; operates tactical support center systems. | Above average learning ability. High degree of electrical and mechanical aptitude. Must pass flight physical and be able to swim. Courses in algebra, trigonometry, physics, electricity. | Radar technician, radio operator. |
| Aviation Fire Control Technician (C) | Maintains and inspects aircraft weapons systems, weapon-control radar, computers, computer sights, gyroscopes, related equipment; air launched guided missile equipment. | Superior electronic, electrical and mechanical aptitude. Training in repair shops or vocational schools and in mathematics. | Instrument person, aircraft electrician, electronics technician, radar computer repairer, TV repairer. |
| Boiler | Operates boilers and fireroom | Strong interest in mechanical | Marine fireman, boiler ship |

| Career Fields | Duties & Responsibilities | Qualifications | Examples of Civilian Jobs |
|---|---|---|---|
| Technician (C) | machinery; transfers, tests and takes inventories of fuel and water, maintains boilers, pumps, associated machinery. | work. Shop courses and practical mathematics. | mechanic, boiler maker, stationary engineer, boiler or heating plant technician. |
| Electronic Warfare Technician (C) | Operates and maintains electronic countermeasures and electronic support measures; associated supporting equipment; evaluates, processes and applies intercept signal data, electronic intelligence reports, and electronic warfare tactics and doctrine to operational needs. | Prolonged attention and mental alertness. Physics, mathematics and courses in radio and electricity helpful, and experience in radio repair. Normal sight and hearing, manual dexterity, and good memory. | Electronics Intelligence Operations Specialist. |
| Fire Control Technician (C) | Operate, test, maintain and repair weapons control systems and telemetering equipment used to compute and resolve factors which influence accuracy of naval guns and missiles. | Perform fine, detailed work. Extensive training in mathematics, electronics, electricity and mechanics. | Radar or electronics technician, test range tracker, instrument repairer, electrician. |
| Gas Turbine Systems Technician (C) | Operates gas turbine engines, main propulsion machinery and related electrical and electronic equipment. | Electrical/electronics repair, blueprint reading, mathematics and physics. | Electronics technician, gas turbine mechanic, power plant operator. |
| Gunner's Mate (C) | Operates and performs maintenance on guided-missile launching systems, rocket launchers, guns, gun mounts; inspects/repairs electrical, electronic, pneumatic, mechanical | Prolonged attention and mental alertness, ability to perform detailed work. High aptitude for electrical and mechanical work. Arithmetic, shop math, electricity, electronics, physics, machine | Electronics technician, electrician, instrument repairer, hydraulics, pneumatic or mechanical technician, small appliance or test equipment repairer. |

| | Duties | Qualifications | Related civilian jobs |
|---|---|---|---|
| **Machinist's Mate (C)** | and hydraulic systems. Operates, maintains and repairs ship's propulsion, auxiliary equipment and outside equipment such as steering, engine, refrigeration/air conditioning, laundry equipment. | shop, welding, mechanical drawing and shopwork. Aptitude for mechanical work. Practical or shop mathematics, machine shop, electricity and physics valuable. | Boiler house repairer, engine maintenance, machinist, marine engineer, turbine operator, engine repairer, air conditioning and refrigeration repairer. |
| **Missile Technician (C)** | Maintains fleet ballistic missiles and support equipment; tests, adjusts, calibrates, operates, repairs support equipment; handles/stows missiles. | High mechanical aptitude and manual dexterity, electricity, electronics, mathematics, and physics. | Rocket engine component ordnance artificer, electronics mechanic. |
| **Operations Specialist (C)** | Operates surveillance and search radar, electronic recognition and identification equipment, controlled approach devices and electronic aids to navigation; serves as plotter and status board keeper. | Prolonged attention and mental alertness. Physics, mathematics and ship courses in radio and electricity helpful. Experience in radio repair is valuable. | Radio operator (aircraft, ship, government service, radio broadcasting), radar equipment supervisor, and control tower operator. |
| **Sonar Technician (C)** | Operates electronic underwater detection and attack apparatus, obtains and interprets information for tactical purposes, maintains and repairs electronic underwater sound detection equipment. | Normal hearing and clear speaking voice. Algebra, geometry, physics, electricity and shopwork desirable. Experience as amateur radio operator. | Oil well sounding device operator, radio operator, inspector of electronic assemblies, electronic technician, electrical repairer. |
| **Aerographer's Mate** | Collects, records, analyzes, meteorological and oceanographic data; enters on appropriate charts; forecasts from visual and | Skills in mathematics, speaking, writing, record keeping and ability to perform repetitive work. | Weather observer, meteorological aide, chart maker, weather chart preparer. |

| Career Fields | Duties & Responsibilities | Qualifications | Examples of Civilian Jobs |
|---|---|---|---|
| | instrumental weather observations. | | |
| **Air Traffic Controller** | Controls air traffic, operates radar air control ashore and afloat; uses radio, light signals; directs aircraft under visual flight and instrument flight conditions; assists in preparation of flight plans. | High degree of accuracy, precision, self-reliance and calmness under stress. Experience in radio broadcasting and good vision. | Control tower operator, radio-telephone operator, flight operations specialist and aircraft log clerk. |
| **Aircrew Survival Equipmentman** | Maintains and packs parachutes, survival equipment, flight and protective clothing, life jackets; tests and services pressure suits. | Must perform extremely careful and accurate work. General shop and sewing desirable. Experience in use and repair of sewing machines. | Parachute packer, inspector, repairer and tester; sailmaker. |
| **Aviation Anti-submarine Warfare Technician** | Performs a wide range of electronic shop operations; performs in-flight maintenance of airborne electronic systems; removes and installs units of anti-submarine warfare equipment; debriefs flight crews; reads and applies service diagrams; performs administrative duties. | Mathematical ability, manual dexterity, ability to do detailed work, a good memory, resourcefulness and curiosity. | Electronics mechanic. |
| **Aviation Boatswain's Mate** | Handles aircraft on carriers; operates, maintains, repairs aviation fueling, defueling, inert gas systems; maintains catapults, arresting gear. | Must have physical strength and manual dexterity. Ability to perform repetitive tasks. | Machinery erector, crane operator, airport service person, gasoline distributor. Fire fighter-crash fire and rescue. |
| **Aviation** | Maintains, adjusts, repairs aircraft | Algebra, trigonometry, physics | Aircraft electrician, airframe and |

| | Duties | Requirements | Civilian Equivalent |
|---|---|---|---|
| **Electrician's Mate** | electrical and instrument systems, plus power generating, lighting, electrical components of aircraft controls. | and shop experience in aircraft electrical work. | power plants mechanic. |
| **Aviation Electronics Technician** | Tests, maintains, repairs aviation electronics equipment including navigation, identification, detection, reconnaissance, special purpose equipment; operates warfare equipment. | High degree of aptitude for electrical, mechanical work and accurate bookkeeping. | Aircraft electrician, radio mechanic, electrical repairer, instrument repairer, electronics technician. |
| **Aviation Machinist's Mate** | Inspects, maintains power plants and related systems and equipment, prepares aircraft for flight, conducts periodic aircraft inspections. | Good learning ability and mechanical aptitude. Machine shop, automobile or aircraft engine work, algebra and geometry. | Airport service person, aircraft engine test mechanic, small appliance repairer. |
| **Aviation Maintenance Administration** | Management and clerical duties in aircraft maintenance offices, plans and schedules maintenance workload, prepares reports and correspondence and analyzes trends of aircraft system and component failures. | Accurate and detailed work, has interest in the aviation maintenance field. Filing and typing. | Shipping, parts, supply room or maintenance clerk, office manager. |
| **Aviation Ordnance** | Loads bombs, torpedoes, rockets, guided missiles; maintains, repairs, inspects aircraft armament, aviation ordnance equipment. | Normal vision and good mechanical aptitude, Algebra, physics and electricity and experience in electrical or mechanical repair. | Gyroscope mechanic, instrument person, ordnanceman, armament inspector. |
| **Aviation Storekeeper** | Receives, stores, issues aviation supplies, spare parts, technical | Ability to keep accurate records and typing helpful. | Accountant, bookkeeper, postal clerk and storekeeper. |

| Career Fields | Duties & Responsibilities | Qualifications | Examples of Civilian Jobs |
|---|---|---|---|
| | aviation items; conducts inventories. | | |
| **Aviation Structural Mechanic** | Maintains and repairs aircraft, air-frame, structural components, hydraulic controls, utility systems, degress systems. | High degree of mechanical aptitude. Metal shop, woodworking, algebra, plane geometry, physics; experience in automobile body work. | Welder, sheet metal repairer, hydraulics technician, radiographer and aircraft plumbing systems mechanic. |
| **Aviation Support Equipment Technician** | Services, tests and performs intermediate level maintenance and repair of gasoline and diesel engines, gas turbine compressor units, power generating equipment, liquid and gaseous oxygen and nitrogen servicing equipment, automotive electrical and air conditioning systems. | High mechanical aptitude, physical strength, machine shop experience, and recordkeeping abilities. | Diesel or gasoline engine, air conditioning, ignition mechanic, hydraulics repairer, auto mechanic. |
| **Boatswain's Mate** | Performs seamanship tasks, operates small boats, stores cargo, handles ropes and lines, directs work of deck force personnel | Must be physically strong. Practical math desirable; algebra, geometry and physics. | Motorboat operator, pier superintendent, able seaman, canvas worker, rigger, cargo wincher, mate, longshore worker. |
| **Builder** | Constructs, maintains, repairs wood, concrete, masonry structures; erects and repairs waterfront structures. | High mechanical aptitude. Carpentry and ship mathematics desirable. Experience with hand and power tools valuable. | Plasterer, roofer, mason, painter, construction worker, carpenter. |
| **Construction Electrician** | Installs, operates, maintains, repairs electrical generating and distribution systems, transformers, switchboards, motors, controllers. | Interest in mechanical and electrical work. Electricity, shop mathematics and physics helpful, ability to work aloft. | Powerhouse or construction electrician, electrical and telephone repairer. |

| | | | |
|---|---|---|---|
| **Construction Mechanic** | Maintains, repairs and overhauls automotive and heavy construction equipment. | High mechanical aptitude. Electrical or machine shop, shop mathematics and physics helpful. Machinist or auto mechanic work. | Automotive or diesel engine mechanic, motor analyst, construction equipment mechanic. |
| **Cryptologic Technician** | | All prospective Cryptologic Technicians (CTs) must have a personal background that will facilitate clearance for special security access; above average speaking and writing ability; good memory; resourcefulness; curiosity; adaptability to detailed work; aptitude for math; record-keeping ability; ability to work well with others and manual dexterity. | |
| **Cryptologic Technician A (Administrative)** | Types messages and correspondence; files; handles classified material; keeps mail logs; prepares correspondence; orders supplies and takes inventory. | Exceptionally good character, speaking and writing ability. | Clerk typist, office manager, intelligence specialists, and keypunch operator. |
| **Cryptologic Technician I (Interpretive)** | Operates technical communications systems equipment; prepares data and reports involving communications material; performs temporary duty aboard submarines and surface units. | Foreign language aptitude. | Interpreter, translator. |
| **Cryptologic Technician M (Maintenance)** | Performs preventive and corrective maintenance on solid-state and electro-mechanical equipment which requires use of test equipment, hand tools and technical publications; repairs and calibrates wide variety of precision electronic test equipment. | Ability to comprehend advanced electronic theory. | Automatic equipment technician; central office repairer, radio mechanic. |

| Career Fields | Duties & Responsibilities | Qualifications | Examples of Civilian Jobs |
|---|---|---|---|
| **Cryptologic Technician O (Communications)** | Prepares messages utilizing teletypewriter equipment; transmits, receives, routes and logs message traffic; maintains message center files/logs and records; controls and operates communications equipment systems including radio receivers, tone-terminal equipment, DC and audio patch boards and communication security devices. | Typing ability, good memory, resourcefulness, curiosity, manual dexterity, aptitude for figures. | Cryptographic machines, telegraphic-teletypewriter operator. |
| **Cryptologic Technician R (Collection)** | Variety of duties associated with operation of teletype and Morse communications systems; operates radio-receiving, direction-finding and technical documents which are predominantly classified. | Good memory, speaking and writing ability. | Morse operator; radio officer. |
| **Cryptologic Technician T (Technical)** | Variety of duties associated with the operation of radio printer and other sophisticated equipment to study signal propagation; operates radio receiving, teletype, recording and related computer equipment; Morse code, security, communications procedures. | Above average speaking and writing ability, good memory. | Cryptoanalyst, digital computer operator; electronic intelligence operation specialist and telegraph typewriter operator. |
| **Data Processing Technician** | Operates data processing equipment including sorters, collators, reproducers, tabulating printers and computers. | High clerical aptitude. Typing, bookkeeping and operating business machines desirable. Experience in mechanical work. | Data typist, keypunch operator, systems analyst, verifier and tabulating machine operator. |

| Rating | Duties | Recommended Subjects | Related Civilian Jobs |
|---|---|---|---|
| **Data Systems Technician** | Maintains electronic digital data systems and equipment; inspects, tests, calibrates, and repairs computers, tape units, digital display equipment, data link terminal sets and related equipment. | Possess high aptitude for detailed mechanical work. Radio, electricity, physics, and mathematics through calculus. | Electrical, electronic repairer or computer repairer, electronics and data systems technician. |
| **Dental Technician** | Assists dental officers in treatment of patients; performs preventive procedures, and various dental department administrative duties. | Hygiene, physiology and chemistry. | Dental technician, dental hygienist, X-ray technician, dentist's assistant, dental laboratory technician. |
| **Disbursing Clerk** | Opens, maintains, closes military pay records; prepares reports and returns on public monies. | Typing, bookkeeping, accounting, business math and office practices. | Paymaster, cashier, statistical or audit clerk, bookkeeper, bookkeeping machine operator, cost accountant. |
| **Electrician's Mate** | Maintains power and lighting equipment, generators, motors, power distribution systems, other electrical equipment; rebuilds electrical equipment. | Aptitude for electrical and mechanical work. Electrical, practical and shop mathematics, and physics. | Electrician, electric motor and electrical equipment repairer, armature winder, radio/TV repairer. |
| **Electronics Technician** | Maintains all electronic equipment used for communications, detection ranging, recognition and countermeasures. | Aptitude for detailed mechanical work. Radio, electricity, physics, algebra, trigonometry and shop valuable. | Electronics technician, radar and radio repairer, instrument and electronics mechanic. |
| **Engineering Aid** | Performs tasks required in construction surveying and drafting. Makes and controls surveys, runs and closes traverses; conducts soil classification and compacting tests. | Algebra, geometry, trigonometry, mechanical drawing and drafting recommended. Experience in road construction useful. | Surveyor, plans drafter, soil analyst. |

| Career Fields | Duties & Responsibilities | Qualifications | Examples of Civilian Jobs |
|---|---|---|---|
| **Engineman** | Operates, maintains, repairs internal combustion engines, main propulsion machinery, refrigeration and assigned auxiliary equipment. | Algebra, geometry and physics helpful. Experience in automotive repair. | Diesel engine operator, diesel mechanic, ignition repairer, small engine mechanic, marine oiler, stationary engineer. |
| **Equipment Operator** | Operates automotive and heavy construction equipment. | Good physical strength and normal color perception. Experience in construction work, auto or electrical shop. | Bulldozer, pile driver, power shovel or motor grader operator, excavation foreman, truck driver. |
| **Hospital Corpsman** | Administers medicines, applies first aid, assists in operating room, nurses sick and injured. | Hygiene, biology, first aid, physiology, chemistry, typing and public speaking. | Practical nurse, medical or X-ray lab technician, nurse administrator. |
| **Hull Maintenance Technician** | Fabricates, installs, repairs shipboard structures, plumbing and piping systems; uses damage control in fire fighting; and nuclear, biological, chemical and radiological defense equipment. | High mechanical aptitude. Sheet metal, foundry, pipefitting, carpentry, mathematics, geometry and chemistry valuable. | Fire fighter, welder, plumber, shipfitter, blacksmith, metallurgical technician. |
| **Illustrator Draftsman** | Designs, sketches, does layout, letters signs, charts and training aids; operates visual presentation equipment; makes mathematical computations for layout and design illustrations. | Previous experience as draftsman, tracer or surveyor valuable. Art, mechanical drawing and blueprint reading valuable. | Structural draftsman, technical illustrator, specification writer, electrical draftsman, geodetic computer, graphic artist. |
| **Intelligence Specialist** | Maintains/uses intelligence files; prepares maps, graphics, mosaics, charts; extracts intelligence information from aerial photos, prepares intelligence reports. | Processing, assimilating, interpreting and presenting data. Typing, filing, drafting, mathematics, geography and photography valuable. | Aerial photographer and intelligence clerk. |

| | Duties | Qualifications | Civilian Occupations |
|---|---|---|---|
| **Interior Communications Electrician** | Maintains, operates all interior communications systems, voice interior communications, alarms, ship's control, plotting, automated propulsion equipment. | High aptitude for electrical work. Electrical shop, practical and shop mathematics, experience in electrical/electronics work desirable. | Powerhouse engineer, ship electrician, station installer, instrument person, electronics or TV technician. |
| **Journalist** | Reports, edits, copyreads news; publishes information about service people and activities through newspapers, magazines, radio and television. | High degree of clerical aptitude. English, journalism, typing and writing experience helpful. | News editor, copyreader, script writer, reporter, rewrite or art layout person, radio & TV announcer, production manager. |
| **Legalman** | Provides administrative services, military justice, claims, administrative law, and legal assistance, serves as court reporter. | Aptitude for detail, ability to express self in writing and orally. Typing, shorthand, English and logic helpful. No speech or hearing difficulties. | Legal assistant, law and contract clerk, title examiner and court reporter, office manager. |
| **Lithographer** | Performs offset lithography and letterpress printing, copy preparation, camera work, assembling and stripping, platemaking, typesetting, presswork and binding. | Work with machinery and chemicals. Printing, physics, chemistry, English and shop mathematics valuable. | Lithographic and plate press operator, bookbinder, engraver, camera operator and photolithographer. |
| **Machinery Repairer** | Maintains assigned equipment to support other ships requiring use of milling machines, boring mills, other machine tools found in machine shop; overhauls and repairs machinery. | Exerience in practical or shop mathematics, machine shop, electricity, mechanical drawing and foundry desirable. | Engine lathe operator, machinist tool clerk, bench machinist, turret and milling machine operator and tool maker. |
| **Master-at-Arms** | Performs investigations, apprehensions, crime prevention, preservation of evidence; | Experience in police, shore patrol or investigative work. Maturity, good vision and hearing. High | Police officer, guard, detective, investigator. |

| Career Fields | Duties & Responsibilities | Qualifications | Examples of Civilian Jobs |
|---|---|---|---|
| | performs duties of beach guard and shore patrol, crowd control and brig operations. | school diploma or equivalent. | |
| Mess Management Specialist | Operates and manages Navy dining facilities and bachelor quarters; estimates quantities and kinds of foodstuffs required; receives, stows and breaks out food items; prepares menus; plans, prepares and serves meals; maintains stock records; conducts inventories; assists medical personnel in inspection for quality; complies with sanitary and hygienic requirements. | Experience or courses in food preparation, dietetics and record keeping helpful. High standards of honesty and cleanliness; good learning ability. | Caterer, cook, steward, chef, mess attendant, motel-hotel services assistant. |
| Mineman | Tests, maintains and repairs mines, components and mine laying equipment. | High mechanical aptitude. Electricity, machine shop work, welding, mechanical drawing and shop mathematics desirable. | Ordnanceman, mine assembler, ammunition foreman, powderman, electrician, harbor patrol. |
| Molder | Operates foundries aboard ship and at shore stations; makes molds and cores, rigs flasks; casts ferrous, non-ferrous and alloy metals; sandblasts castings and pours bearings. | Foundry, machine shop, practical mathematics. | Foundry supervisor, furnace operator, melter, molder, core maker, heat treater, temperer. |
| Musician | Provides music for military ceremonies, religious services, concerts, parades, various recreational activities; plays one or more musical instruments. | Proficiency on standard band or orchestral instruments. | Music teacher, instrument musician, orchestra leader, music arranger, instrument repairer, music librarian, arranger. |

| | | | |
|---|---|---|---|
| **Ocean Systems Technician** | Operates special electronic equipment to interpret and document oceanographic data; operates related equipment such as tape recorders; interprets data; prepares and maintains visual displays of data; converts data into formats for statistical study. | Normal hearing, vision and color perception; above average learning ability; ability to perform detailed and repetitive work, to work harmoniously with others, and with numbers; and qualified for secret security clearance. | Computer-peripheral equipment operator or electronics technician. |
| **Opticalman** | Maintains binoculars, sextants, optical gunsights, turret and marine periscopes. | Close, exact and painstaking detail work, physics, shop mathematics and machine shop helpful; experience in optical or camera manufacturing. | Lens grinder, jewelry stone cutter, tool inspector, optical maker, inspector, optical tooling specialist, camera repairer and locksmith. |
| **Patternmaker** | Makes wood, plaster, metal patterns, core boxes, flasks used by molders in Navy foundries. | Exacting, precise work; woodshop, foundry, mechanical drawing, shop and practical mathematics. | Template maker, industrial arts teacher, layout person, pattern-maker, form builder. |
| **Personnelman** | Performs enlisted personnel administration duties in manpower utilization, maintains service records, personnel accounting, educational services, classifies personnel and jobs. | Ability to deal with people, typing, public speaking, office practices, personnel work and counseling helpful. | Employment manager, personnel clerk, vocational adviser, clerk typist, job or organizational analyst. |
| **Photographer's Mate** | Operates, maintains and repairs cameras for ground and aerial photographic work. | Normal color perception; physics and chemistry desirable. | Photographer, camera repairer, aerial photographer. |
| **Postal Clerk** | Processes mail, sells stamps and money orders, maintains mail directories and handles correspondence concerning postal operations. | Bookkeeping, accounting, business math, typing and office practices. | Parcel post or mail clerk, mail room manager, stock clerk, cashier. |

| Career Fields | Duties & Responsibilities | Qualifications | Examples of Civilian Jobs |
|---|---|---|---|
| Quartermaster | Performs navigation of ships, steering, lookout supervision, ship-control, bridge watch duties visual communication and maintenance of navigation aids. | Good vision and hearing and ability to express oneself clearly in writing and speaking. Public speaking, grammar, geometry and physics helpful. | Barge, motorboat, yacht captain, quartermaster, harbor pilot aboard merchant ships. |
| Radioman | Operates communication, transmission, reception, and terminal equipment; transmits, receives and processes all forms of military record and voice communications. | Good hearing and manual dexterity. Mathematics, physics and electricity desirable. Experience as amateur radio operator helpful. | Telegrapher, radio dispatcher, radio/telephone operator, news copywriter. |
| Religious Program Specialist | Assists in management of religious programs and facilities; trains volunteers; supervises the offices of chaplains; performs administrative duties. | Relates easily with people. Basic English, business math, typing, graphics and audiovisual familiarization useful. | Church business manager or administrator. |
| Ship's Serviceman | Operates and manages ship's store activities afloat and ashore, including barber, cobbler, tailor, laundry, dry cleaning, commissaries, retail stores. | Shoe repairing, barbering, tailoring, merchandising and retailing, accounting, bookkeeping, business math and English helpful. | Barber, laundry and dry cleaner, retail store manager, sales clerk, tailor, and shoe repairer. |
| Signalman | Sends and receives messages by flashing light, semaphore and flag hoist; handles, routes and files messages; codes and decodes message headings; operates voice radio and maintains visual sight equipment. | Good vision and hearing, ability to express oneself clearly in writing and speaking. | Third mate, signalman, deck cadet, harbor police, small boat operator. |

| | Duties | Qualifications | Related Civilian Occupations |
|---|---|---|---|
| **Steelworker** | Fabricates, erects and dismantles pre-engineered structures, steel bridges and other structures. Lays out and fabricates steel and sheet metal; welds. | Physical strength, stamina and ability to work aloft. Sheet metal, machine shop, foundry experience desirable. | Rigger, shipfitter, structural steelworker, salvage engineer, steel fabricator, welder, sheet metal technician. |
| **Storekeeper** | Orders, receives, stores, inventories and issues clothing, foodstuffs, mechanical equipment and other items. | Typing, bookkeeping, accounting, commercial math, general business studies and English. | Sales or shipping clerk, warehouse worker, buyer, invoice control clerk, purchasing agent, accountant. |
| **Torpedoman's Mate** | Maintains and overhauls torpedoes and depth charges; maintains and repairs ordnance launching equipment; launches and recovers torpedoes. | High mechanical and electrical aptitude. Electricity, machine shop, welding, mechanical drawing and ship mathematics desirable. | Ordnance supervisor, gyroscope assembly supervisor, instrument mechanic, electronics technician, motor/office machine repairer. |
| **Utilitiesman** | Installs, maintains, repairs and codes plumbing, heating systems, steam, compressed air, fuel storage, collection and disposal facilities and water purification units. | High mechanical aptitude. Apprentice training in plumbing and related fields, mathematics helpful. | Stationary engineering assistant, plumber, pipe fitter, water plant or boiler operator, boiler house supervisor. |
| **Yeoman** | Clerical and secretarial, typing, filing, operating office and duplicating equipment, preparing and routing correspondence and reports, maintaining records and official publications. | Same qualifications required of secretaries and typists in private industry; English, business subjects, stenography and typewriting helpful. | Office manager, secretary, general office clerk, administrative assistant. |

## Air Force Occupations

| Career Fields | Duties & Responsibilities | Qualifications | Examples of Civilian Jobs |
|---|---|---|---|
| **Accounting, Finance and Auditing** | Prepares documents required to account for and disburse funds, including budgeting allocation, disbursing, auditing and preparing cost analysis records. | Dexterity in the operation of business machines. Typing, mathematics, statistics and accounting desirable. High administrative aptitude mandatory. | Public accountant, auditor, bookkeeper, budget clerk and paymaster. |
| **Administration** | Prepares correspondence, statistical summaries, arranges priority and distribution systems, maintains files, prepares and consolidates reports and arranges for graphic presentation. | Business, English, typing and mathematics courses desirable. | Clerk typist, file secretary, stenographer, receptionist. |
| **Aircraft Maintenance** | Performs the mechanical functions of maintenance, repair, and modification of helicopters, turbo-prop, reciprocating engine and jet aircraft. | Considerable mechanical or electrical aptitude and manual dexterity. Physics, hydraulics, electronics, mechanics and mathematics desirable. | Aircraft mechanic, airframe inspector. |
| **Aircraft Systems Maintenance** | Performs maintenance of aircraft accessory systems, propulsion systems, fabrication of metal and fabric materials used in aircraft structural repair, and inspection and preservation of aircraft parts and materials. | Electrical or mechanical aptitude and manual dexterity. Electronics, mathematics, hydraulics, mechanics, chemistry, metal-working and mechanical drafting desirable. | Aircraft mechanic, aircraft electrician, sheet metal worker, welder, machinist. |
| **Aircrew Operations** | Primary duties require frequent and regular flights. Inflight | High electrical and mechanical aptitude, manual dexterity, | Aircraft mechanic, electrician, hydraulic tester, oxygen systems |

| | | |
|---|---|---|
| Refueling Operator performs duties associated with inflight refueling of aircraft; Defensive Aerial Gunner is a B-52 integrated crewmember with responsibility for defense of the aircraft; Aircraft Loadmaster supervises loading of cargo and passengers and operates aircraft equipment; Pararescue/Recovery personnel perform aircrew protection skills; and Flight Engineers ensure mechanical condition of the aircraft and monitor inflight aircraft systems. | normal vision and good physical condition. Mathematics, physics, general science, English, typing, computer principles, and shopwork desirable. | tester, cargo handler, dispatcher and shipping clerk depending upon the area in which training and experience is received. No civilian job covers some aspects of this field. |
| **Aircrew Protection** Peforms functions involved in the instruction of aircrew and other designated personnel on the principles, procedures, and techniques of global survival. This includes life support equipment, recovery, evasion, captivity, resistance to exploitation and escape. | Good physical condition required; knowledge of pioneering and woodsman activities helpful. Courses in communications, science, and education desirable. | No civilian job covers the scope of the jobs in this career field, but a related job is that of hunting or fishing guide. |
| **Avionic Systems** Installs, maintains and repairs airborne bomb navigation, fire control, weapon control, automatic flight control systems, radio and navigation equipment and maintains associated test and precision measurement equipment. | Electronic aptitude, manual dexterity and normal vision. Mathematics, physics, chemistry, electronics and trigonometry desirable. | Radar, television and precision instrument maintenance. |

| Career Fields | Duties & Responsibilities | Qualifications | Examples of Civilian Jobs |
|---|---|---|---|
| **Band** | Plays musical instruments in concert bands and orchestras, repairs and maintains instruments, vocalist, performs as drum major, arranges music and maintains music libraries. | Knowledge of rudiments of music, elementary theory of music and orchestration desirable. | Orchestrator, music librarian, music teacher, instrumental musician. |
| **Command Control Systems Operations** | Performs functions involved in aerospace surveillance and aerospace vehicle detection, including missile warning systems, controlling, and plotting. Includes control tower and airways operation; ground-controlled approach procedures; operation of all types of ground radar and related communications equipment, except weather equipment. | Good physical and emotional condition required. Courses in typing, mathematics, business machines, communication, English, and science desirable. | |
| **Communication-Electronics Systems** | Installs, modifies, maintains, repairs and overhauls airborne and ground television equipment, high speed general and special purpose data processing equipment, automatic communications and cryptographic machines systems, teletypewriter, teleautographic equipment, telecommunications systems control and associated electronic test equipment. | Basic knowledge of electronic theory. Mathematics and physics desirable. Normal color vision mandatory. | Communications, electronics technician, radio and television repairer, meteorological and teletype equipment repairer. |

| | | |
|---|---|---|
| **Control Systems Operations** | Operates control towers, directs aircraft landings with radar landing control equipmment; operates ground radar equipment, aircraft control centers, airborne radar equipment, space tracking and missile warning systems. | Equipment dexterity, clear voice and speech ability and excellent vision. English desirable. | Aircraft log clerk, airport control operator and air traffic controller. |
| **Dental** | Operates dental facilities and provides paraprofessional dental care; preventive dental services, treatment of oral tissues and fabricates prosthetic devices. | Knowledge of oral and dental anatomy; biology and chemistry desirable. | Dental hygienist, dental assistant. |
| **Education and Training** | Conducts formal classes of instruction, uses training aids, develops material for various courses of instruction; teaches classes in general academic subjects and military matters, and administers educational programs. | English composition and speech desirable. | Vocational training instructor, counselor, educational consultant, or administrator. |
| **Fire Protection** | Operates fire fighting equipment, prevents and extinguishes aircraft and structural fires; rescues and renders first aid; maintains fire fighting and fire prevention equipment. | Good physical condition, no allergies to oil and fire extinguishing solutions; general science and chemistry desirable. | Fire chief, fire extinguisher service person, fire fighter, fire marshal and fire department person. |
| **Fuels** | Receives, stores, dispenses, tests and inspects propellants, petroleum fuels and products. | Chemistry, math, and general science desirable. | Petroleum industry supervisor and bulk plant manager. |
| **Geodetic** | Procures, compiles, computes and uses topographic, | Ability to use precision instruments required in measuring | Cartographer, topographical drafter, mapmaker, and |

| Career Fields | Duties & Responsibilities | Qualifications | Examples of Civilian Jobs |
|---|---|---|---|
| | photogrammetric, and cartographic data in preparing aeronautical charts, topographic maps and target folders. | and drafting; algebra, geometry, trigonometry and physics necessary. | advertising layout person. |
| **Information Systems Operator** | Operates radio and wire communications systems, automatic digital switching equipment, cryptographic devices, airborne and ground electronic countermeasures equipment, all kinds of communication equipment, and the management of radio frequencies. Collects, processes, records, prepares and submits data for various automated systems, analyzes design, programs and operates computer systems. | Knowledge of telecommunications functions and operations of electronic communications equipment. Typing or keyboard experience, and clear speaking voice desirable in many specialties. Business math, algebra and geometry desirable. | Central office operator (telephone and telegraph), cryptographer, radio operator, telephone supervisor and photo-radio operator. Card-tape converter or computer operator, data typist, data processing control clerk, high-speed printer operator, programmer. |
| **Intelligence** | Collects, produces and disseminates data of strategic, tactical or technical value from an intelligence viewpoint. Maintains information security. | Knowledge of techniques of evaluation, analysis, interpretation and reporting, foreign languages, English composition, photography, mathematics and typing desirable. | Cryptoanalyst, drafts person, interpreter, investigator, statistician, radio operator and translator. |
| **Intricate Equipment Maintenance** | Overhauls and modifies photographic equipment; work with fine precision tools, testing devices and schematic drawings. | Considerable mechanical ability and manual dexterity; algebra and physics desirable. Must have normal color vision. | Camera repairer, statistical machine and medical equipment service person. |

| | | | |
|---|---|---|---|
| **Legal** | Takes and transcribes verbal recordings of legal proceedings, uses stenomask; performs office administrative tasks; processes claims. | Knowledge of stenomask, typewriter, legal terminology, military processing of claims; English grammar and composition; ability to speak clearly and distinctly. | Law librarian, court clerk and shorthand reporter. |
| **Management Analysis** | Collects, processes, records, controls, analyzes, and interprets special and recurring reports, statistical data and other information. | Knowledge of business statistics, mathematics, accounting and English desirable. Completion of high school or GED equivalent mandatory. | Statistical, accounting and budget clerk. |
| **Mechanical/ Electrical** | Performs installation, operation, maintenance and repairs of base direct support systems and equipment. | Physics, mathematics, blueprint reading and electricity. | Elevator repairer, electrician, lineman, powerhouse repairer, diesel mechanic, pipefitter, steamfitter and heating and ventilating worker. |
| **Medical** | Operates medical facilities, works with professional medical staff as they provide care and treatment. May specialize in such medical services as nuclear medicine, cardiopulmonary techniques, physical and occupational therapy, orthopedic appliances, medical laboratory and medical administrative services. | Knowledge of first aid, ability to help professional medical personnel; anatomy, biology, zoology. High school algebra and chemistry desirable in most specialties and are mandatory requirements for some. | X-ray and medical record technician, medical laboratory and pharmacist assistant. Respiratory therapy technician and surgical technologist. |
| **Missile Systems Maintenance** | Assemble, transport, install, maintain, inspect, modify, check out and repair missiles, missile airframes and subsystems and remotely piloted vehicles. | Mechanical or electrical aptitude and manual dexterity. Mathematics and physics desirable. Normal color vision mandatory. | Missile facilities repairer, electronics mechanic, and aircraft mechanical/electrical system repairer. Mechanical inspector, mechanical engineer, aircraft |

| Career Fields | Duties & Responsibilities | Qualifications | Examples of Civilian Jobs |
|---|---|---|---|
| | Inspects, maintains, repairs, calibrates, and modifies missile facilities support systems, liquid propellant systems, test equipment, and related missile and remotely piloted vehicle sub-systems. | | mechanic and pneumatic tester and mechanic. |
| **Missile Maintenance** | Performs missile engine installation, maintenance and repair; maintenance, repair and modification of missile airframes, subsystems and associated aerospace ground equipment. | Mechanical aptitude and manual dexterity. Mathematics and physics desirable. Normal color vision mandatory. | Mechanical inspector, mechanical engineer, aricraft mechanic, and pneumatic tester and mechanic. |
| **Morale, Welfare and Recreation** | Conducts physical conditioning, coaches sports programs, administers recreation, entertainment, sports and club activities. | Good muscular coordination; English, business math and physical education desirable. | Athletic or playground director, physical education instructor and manager of a recreational establishment. |
| **Motor Vehicle Maintenance** | Overhauls and maintains powered ground vehicles and mechanical equipment for transporting personnel and supplies. | Machine shop, mathematics and training in the use of tools and blueprints helpful. | Automobile accessories installer, automobile repairer, bus mechanic, carburetor person, automotive electrician and truck mechanic. |
| **Munitions and Weapons Maintenance** | Maintains and repairs aircraft armament; assembles, maintains, loads, unloads and stores munitions and nuclear weapons; disposes of bombs, missiles and rockets and operates detection instruments. | Mechanical or electrical aptitude, manual dexterity, normal color vision and depth perception. Mathematics, mechanics, and physics desirable. | Aircraft armament mechanic, armorer, ammunition inspector, munitions handler. |

| | Duties | Qualifications | Related civilian jobs |
|---|---|---|---|
| **Personnel** | Interviews, classifies, selects career jobs for airmen on the basis of qualifications and requirements of the Air Force; administers aptitude, performance tests; administers personnel quality control programs; performs counseling, educational and administrative functions. | English composition and speech. Operation of simple data processing equipment and typing ability desirable. | Employment or personnel clerk, special services supervisor, personnel service manager, personnel supervisor, counselor. |
| **Public Affairs** | Interviews people; reports news; composes, proofreads, writes and edits news copy; provides public affairs advice. | High general aptitude and completion of high school or GED equivalency mandatory. English grammar and composition, speech, journalism, drama, radio/television, history, or political science desirable. | Reporter, copy reader, historian, public relations representative, editorial assistant, broadcast or program director, announcer. |
| **Reprographic** | Operates and maintains reproduction equipment used in the graphic arts, performs hand and machine composition and binding operations. | Mechanical ability and dexterity; courses in chemistry and shop mechanics desirable. | Lithographic press, fold machine, offset and webpress, perforating machine or duplicating machine operator; bookbinder; photolithographer; photoengraver. |
| **Safety** | Performs functions related to the conduct of both safety and disaster preparedness programs. Conducts safety programs, surveys areas and activities to eliminate hazards, analyzes accident causes and trends. Trains personnel to accomplish the primary mission under the handicaps imposed by | Knowledge of industrial hygiene, safety education, safety psychology, and blue-print interpretation. Typing, English, public speaking, mathematics, and science desirable. | Safety inspector and instructor. |

| Career Fields | Duties & Responsibilities | Qualifications | Examples of Civilian Jobs |
| --- | --- | --- | --- |
| | enemy attack and by acts of man and nature. | | |
| **Sanitation** | Operates and maintains water and waste processing plant systems and equipment and performs pest and rodent control functions. | Physics, biology, chemistry and blueprint reading valuable. | Purification plant operator, sanitary inspector, exterminator and entomologist. |
| **Security Police** | Provide security for classified information and material, enforce law and order, control traffic, and protect lives and property, organize as local ground defense forces. | Good physical condition, vision and hearing; civics and social sciences desirable. | Guard, police inspector, police officer, and superintendent of police. |
| **Special Investigations/ Counterintelligence** | Investigates violations of the Uniform Code of Military Justice and applicable federal statutes, investigates conditions pertaining to sabotage, espionage treason, sedition and security. | Knowledge of law enforcement and security regulations, good physical condition, hearing and vision; civics, social sciences, accounting and foreign language desirable. | Detective, chief of detectives, detective sergeant and investigator. |
| **Structural/ Pavements** | Constructs and maintains structural facilities and pavement area; maintains pavements, railroads and soil bases; performs erosion control; operates heavy equipment; performs site development, general maintenance, cost and real property accounting, work control functions and metal fabricating. | Blueprint reading, mechanical drawing, mathematics, physics, and chemistry. | Plumber, bricklayer, carpenter, stonemason, painter, construction worker, welder and sheet metal worker. |

| | | | |
|---|---|---|---|
| **Supply** | Designs, develops, analyzes and operates supply systems including supply data systems; responsible for computation, operation and management of material facilities; equipment review and validation; records maintenance, inventory and distribution control; budget computation and financial plans. | Accounting and business administration. | Junior accountant, machine records section supervisor, receiving, shipping and stock clerk. |
| **Services and Food Services** | Supervises and operates sales stores, laundry/dry cleaning facilities, commissaries and meat processing. Cooks and bakes. | Chemistry, management, marketing, manual dexterity and business mathematics. | Department manager, retail general merchandise manager, meat cutter, butcher, chef and pastry cook. |
| **Training Devices** | Installs, maintains, repairs, modifies and operates training devices such as cockpit procedures trainers, flight and mission trainers and simulators, navigation and tactics training devices, visual training devices, missile crew procedures trainers, and offensive and defensive systems trainers. | Knowledge of electricity, mathematics, blueprint reading and physics desirable. | Link trainer, instructor, radio mechanic. |
| **Transportation** | Ensures service, efficiency and economy of transportation of supplies and personnel by aircraft, train, motor vehicle and ship. | Driver training, operation of office machines and business math. | Cargo handler, motor vehicle dispatcher, shipping or traffic rate clerk, trailer truck driver and ticket agent. |
| **Visual Information** | Operates aerial and ground cameras, motion picture and other photographic equipment; processes photographs and film, | Considerable dexterity on small precision equipment; excellent eyesight. Mathematics, physics, chemistry, public speaking, | Photographer, darkroom technician, film editor, aerial commercial photographer, photograph finisher, sound mixer |

| Career Fields | Duties & Responsibilities | Qualifications | Examples of Civilian Jobs |
|---|---|---|---|
| | edits motion pictures, performs photographic instrumentation functions, and operates airborne, field and precision processing laboratories. | commercial art, drafting, photography, drama, communicative arts, and computer science desirable. | and motion picture camera operator. |
| **Wire Communications Systems Maintenance** | Installs and maintains wire communications equipment and systems. Installs, repairs and maintains telephone and telegraph land line systems, telephone equipment, antenna support systems, key systems, telephone switching equipment, missile communications control systems and electronic switching equipment. | Mechanical/electronic aptitude and manual dexterity; physics and mathematics desirable. Normal color perception mandatory. Physical ability to climb required in some specialties. | Cable splicer, central office repairer, line installer and inspector, teletype and central office manual equipment repairer. |
| **Weather** | Collects, records and analyzes meteorological data; makes visual and instrument weather observations. Forecasts immediate and long-range weather conditions. | Visual acuity correctable to 20/20; physics, mathematics and geography desirable. | Meteorologist, weather forecaster and weather observer. |

## Marine Corps Occupations

Career fields are listed alphabetically, beginning with those combat-related groups closed to women (indicated by "C").

| Career Fields | Duties & Responsibilities | Qualifications | Examples of Civilian Jobs |
|---|---|---|---|
| Field Artillery (C) | Maintains heavy mortars, 175mm, 155mm, 8-inch and 105mm guns; and self-propelled 175mm, 8-inch and 105mm guns. | Mathematical reasoning, mechanical aptitude, good vision and stamina; mechanics, electricity, meteorology and mathematics. | Surveyor, geodetic computer, meteorologist, radio operator, recording engineer and ordnance inspector. |
| Infantry (C) | Performs as rifleman, machine gunner, or grenadier; infantry unit leader, supervises training and operations of infantry units. | Verbal and mathematical reasoning, good vision and stamina; general mathematics, mechanical drafting, geography and mechanical drawing. | Firearms assembler, gunsmith, policeman, immigration inspector and plant security policeman. |
| Tank and Amphibian Tractor (C) | Performs as driver, gunner and loader in tanks, armored amphibious tractors. | Mechanical ability and stamina; auto mechanics, machine shop, electricity and mechanical drawing. | Automotive mechanic, bulldozer operator or repairman, caterpillar repairman, armament machinist-mechanic and gunsmith assistant. |
| Aircraft Maintenance | Performs the mechanical functions of maintenance, repair and modification of Marine air and ground support equipment. | Mechanical or electrical aptitude with manual dexterity; shop mathematics desirable. | Aircraft mechanic, electrician or hydraulics specialist, aviation machinist, sheet metal worker, aircraft instrument maker, repairer. |
| Airfield Services | Maintains aircraft log books, publications and flight operations records; prepares reports and schedules; installs and repairs aircraft launching and recovery equipment. | Typing, geography and mechanical drawing useful. | Airplane dispatch clerk, flight dispatcher, timekeeper and airport crash truck driver. |

| Career Fields | Duties & Responsibilities | Qualifications | Examples of Civilian Jobs |
|---|---|---|---|
| Air Traffic Control and Enlisted Flight Crews/Air Support/Anti-Air Warfare | Operates airfield control tower and radio-radar air traffic control systems; serves as navigator, radio and radar operator, and intercept controller anti-air warfare missile batteryman. | Clear speaking voice, good hearing and better than average eyesight; speech, mathematics and electricity; and experience as a ham radio operator helpful. | Airport control tower, or flight radio operator, navigator, instrument-landing truck operator, radio or television studio engineer. |
| Ammunition and Explosives Ordnance Disposal | Inspects, issues and supervises storage of ammunition and explosives; locates, disarms, detonates or salvages unexploded bombs. | Mechanics, general science, and chemistry useful. | Firearms and ammunition proof director, ordnance technician (government), powder and explosives inspector. |
| Auditing, Finance and Accounting | Prepares and audits personnel pay records, processes public vouchers and administers and audits unit fiscal accounts. | Computation work and attention to detail; typing, bookkeeping, office machines and mathematics useful. | Payroll or cost clerk, bookkeeper, cashier, bank teller, accounting and audit clerk and accountant. |
| Aviation Ordnance | Maintains and repairs aircraft armament systems, gun pods, machine guns, bomb racks and rocket/missile launcher equipment; assembles and loads bombs, rockets, guns and missiles; handles and stores aviation type munitions. | Electricity, hydraulics and mechanics shop courses useful. | Firearms assembler, gunsmith, armament mechanic and aircraft accessories repairer. |
| Avionics | Installs and repairs aircraft electrical, communications/ navigation and fire control equipment and air launched guided missiles; serves as | Mathematics and shop courses in electricity, hydraulics and electronics useful. | Radio and television or electrical instrument repairer, communications, electrical or electronics engineer and radio operator. |

| | Duties | Helpful Courses | Related Careers |
|---|---|---|---|
| **Data/ Communications/ Maintenance** | electrician and instrument repairman. Repairs and calibrates related test equipment. Installs, inspects and repairs telephone, teletype and cryptographic equipment and cables, calibrates precision electronic, mechanical, dimensional and optical test instruments. | Mathematics, electricity and blueprint reading courses helpful. | Telephone installer and trouble shooter, radio repairer, cable splicer, and office machine service person. |
| **Data Systems** | Operates and programs data processing equipment. | Clerical aptitude, manual dexterity and eye-hand coordination, mathematics, accounting and English useful. | Computer operator, programmer and data control coordinator. |
| **Drafting, Surveying and Mapping** | Makes architectural and mechanical drawings, prepares military maps, makes topographic maps, creates or copies articles or illustrative materials. | Mathematics, mechanical drawing and drafting, geography and commercial art helpful. | Architectural or mechanical drafting, surveyor or cartographer, geodetic computer, illustrator and commercial artist. |
| **Electronics Maintenance** | Installs, tests and repairs air-search radar, radio, radio relay, missile fire control and guidance systems. | Electronics, mathematics, electricity, and blueprint reading useful. | Radio and television repair, radio engineer, electrical instrument repairer, recording communications, and electrical engineer. |
| **Engineer/ Construction Equipment and Shore Party** | Performs metal-working, operation and maintenance of fuel storage area, heavy engineering and pioneer equipment, construction and repair of military facilities. | Automotive mechanics, sheet metal working machine shop, carpentry and mechanical drafting useful. | Sheet metal worker, engineering equipment mechanic, carpenter, road machinery operator, rigger and construction superintendent. |

| Career Fields | Duties & Responsibilities | Qualifications | Examples of Civilian Jobs |
|---|---|---|---|
| **Food Service** | Performs as cook, baker or meat cutter. | Hygiene, biology, chemistry, home economics and bookkeeping courses helpful. | Cook, chef, baker, meat cutter or butcher, caterer, executive chef, dietician and restaurant manager. |
| **Intelligence** | Collects, records, evaluates and interprets information, makes detailed study of aerial photographs, conducts interrogations in foreign languages, translates written material and interprets conversations. | Geography, history, government, economics, English, foreign languages, typing, mechanical drafting and mathematics beneficial. | Investigator, research worker, intelligence analyst (government), map drafter, cartographic aide and records analyst. |
| **Legal Services** | Prepares legal documents, operates stenotype machines. | Manual dexterity; English. | Law clerk, court reporter, chief clerk and stenotype operator. |
| **Logistics** | Performs administrative duties involving the supply, quartering movement and transport of Marine units by land, sea and air. | Clerical aptitude, knowledge of verbal and math reasoning; operate office machines and read maps. | Inventory or shipping clerk, pier superintendent, stock control clerk or supervisor and warehouse manager. |
| **Marine Corps Exchange** | Keeps and audits books and financial records, performs sales and merchandise stock control duties. | Typing, bookkeeping, business arithmetic, office machines and accounting useful. | Salesman, stock control supervisor, buyer, bookkeeper, accounting clerk, accountant and auditor. |
| **Military Police and Corrections** | Enforces military orders, guards military and war prisoners and controls traffic. | Sociology and athletics helpful. | Policeman, ballistics expert and investigator. |
| **Motor Transport** | Performs auto mechanics and body repair, motor vehicle and amphibian truck operation. | Automotive mechanics, machine shop, electricity and blueprint reading useful. | Mechanic or automobile body, electrical systems repairer, truck driver, motor vehicle dispatcher, and motor transport. |

| Musician | Performs in Marine Corps Band, unit bands, and drum and bugle corps; repairs musical instruments. | Music experience as a member of a high school band or orchestra. | Musician, music librarian, music teacher, bandmaster, orchestra or music director and musical instrument repairer. |
|---|---|---|---|
| Nuclear, Biological and Chemical | Performs routine duties incident to applying detection, emergency and decontamination measures to gassed or radioactive areas. Inspects and performs preventive maintenance on chemical warfare protection equipment. | Must not have any known hypersensitivity to the wearing of protective clothing; be emotionally stable; biology and chemistry background beneficial. | Laboratory assistant (nuclear, biological or chemical), exterminator, decontaminator. |
| Operational Communications | Lays communication wire; installs and operates radio, radio telegraph and radio relay equipment; encodes and decodes messages. | Mathematics, typing, electricity and electronics useful. | Radio operator, telephone line-person, radio broadcasting traffic manager and communications engineer. |
| Ordnance | Inspects, maintains and repairs infantry, artillery, and anti-aircraft weapons; fire control optical instruments; operates machine tools or modifies metal parts. | Mathematics, mechanics, machine shop and blueprint reading; welding and heat treatment of metal and electricity. | Armament mechanic, gunsmith, time-recording equipment service person, electrician, optical instrument inspector and electrical engineer. |
| Personnel and Administration | Performs as personnel classification clerk, administrative specialist and postal clerk. | Reasoning and verbal ability, and clerical aptitude. English composition, typing, shorthand and social studies helpful. | Secretary-typist, vocational advisor, employment interviewer, manager, office manager, job analyst and postal clerk. |
| Printing and Reproduction | Performs letterpress and lithographic offset printing; sets type; operates linotype machines, presses, process cameras and bookbinding equipment. | General mathematics, printing and other graphic arts useful. | Printing compositor, linotype operator, photolithographer, press operator, printing bookbinder, printing plant makeup worker and proofreader. |

| Career Fields | Duties & Responsibilities | Qualifications | Examples of Civilian Jobs |
|---|---|---|---|
| **Public Affairs** | Gathers material for, writes and edits news stories, historical reports, gathers material for, prepares and edits radio and television broadcast scripts. | English grammar and composition, typing, speech and journalism courses helpful. | News reporter-correspondent, columnist, copyreader, copy or news editor, radio-television announcer and script writer. |
| **Signals Intelligence/ Ground Electronic Warfare** | Performs routine duties incident to collecting, translating, recording and disseminating information associated with military plans and operations. | English composition, geography and mathematics beneficial. | Radio intelligence operator, intelligence analyst, investigator and records analyst. |
| **Supply Administration and Operations** | Administration procurement, subsistence, packaging and warehousing; requisitions, purchases, receipts, accounts, classifies, stores, issues, sells, packages, preserves and inspects new scrap, salvage, waste material, supplies and equipment. | Typing, bookkeeping, office machine operation and commercial subjects helpful. | Shipping, receiving, stock and inventory clerk, stock control supervisor, warehouse manager, parts and purchasing agent. |
| **Training and Audiovisual Support** | Operates still, motion picture, and aerial cameras; develops films and prints photographs; repairs cameras and edits motion picture films; performs as illustrator or drafter. | Mathematics, chemistry and shop course in electricity; normal color perception desirable. | Commercial illustrator, photographer, cinematographer, copy camera operator, motion picture film editor, camera and instrument repairer. |
| **Transportation** | Handles cargo and transacts business of freight shipping and receiving and passenger transportation. | Typing, bookkeeping, business arithmetic, office machine operation, and commercial subjects beneficial. | Shipping clerk, cargo handler, traffic rate clerk, freight traffic, passenger and railroad station agent. |

| Career Fields | Duties & Responsibilities | Qualifications | Examples of Civilian Jobs |
|---|---|---|---|
| Utilites | Installs, operates, and maintains electrical, water supply, heating, plumbing, sewage, refrigeration, hygiene and air conditioning equipment. | Mechanical aptitude and manual dexterity important; vocational school shop course in industrial arts and crafts beneficial. | Electrician, plumber, steam fitter, refrigeration mechanic, electric motor repairer and stationary engineer. |
| Weather Service | Collects, records and analyzes meteorological data; makes visual and instrumental observations and enters them on appropriate charts; forecasts short, intermediate and long range weather conditions. | Visual acuity correctable to 20/20, normal color perception; mathematics desirable, meteorology and astronomy helpful. | Meteorologist, weather forecaster and observer. |

## Coast Guard Occupations

| Career Fields | Duties & Responsibilities | Qualifications | Examples of Civilian Jobs |
|---|---|---|---|
| Aviation Electrician's Mate | Maintains, adjusts, repairs aircraft electrical and instrument systems, plus power generating, lighting, electrical components of aircraft controls. | Algebra, trigonometry, physics and shop experience in electrical work. | Aircraft electrician, electrician, substation operator, instrument inspector, electrical consultant. |
| Aviation Electronics Technician | Tests, maintains, repairs aviation electronics equipment including navigation, identification, detection, reconnaissance, special purpose equipment; operates warfare equipment. | High degree of aptitude for electrical and mechanical work. Algebra, trigonometry, physics, electricity, radio and mechanics. | Aircraft electrician, radio mechanic, electronics technician, radar repairman/technician, TV repairer. |
| Aviation Machinist's Mate | Inspects, maintains power plants and related systems and equipment, prepares aircraft for flight, conducts periodic aircraft inspections | Good learning ability and mechanical aptitude. Machine shop, automobile or aircraft engine work, algebra and geometry. | Airport service person, aircraft engine test mechanic, small appliance repairer, mechanic, machinist, flight engineer. |

| Career Fields | Duties & Responsibilities | Qualifications | Examples of Civilian Jobs |
|---|---|---|---|
| **Aviation Structural Mechanic** | Maintains and repairs aircraft, airframe, structural components, hydraulic controls, utility systems, degree systems. | High degree of mechanical aptitude. Metal shop, woodworking, algebra, plane geometry, physics; experience in automobile body work. | Welder, sheet metal repairer, hydraulics technician, aircraft repairer. |
| **Aviation Survivalman** | Maintains and packs parachutes, survival equipment, flight and protective clothing, life jackets; tests and services pressure suits. Cares for search and rescue equipment, pyrotectnics and station small arms. | Must perform extremely careful and accurate work. General shop and math, sewing desirable. Experience in use and repair of sewing machines. | Parachute packer, inspector, repairer and tester, sailmaker, ammunition foreman, rescue gear specialist. |
| **Boatswain's Mate** | Performs seamanship tasks, operates small boats, stores cargo, handles ropes and lines, directs work of deck force personnel. | Must be physically strong. Practical math desirable; algebra, geometry and physics. | Motorboat operator, pier superintendent, able seaman, canvas worker, rigger, cargo wincher, mate, longshore worker, marina operator, heavy equipment operator. |
| **Damage Controlman** | Fabricates, installs, repairs shipboard structures, plumbing and piping systems; uses damage control in fire fighting, and nuclear, biological, chemical and radiolgoical defense equipment. Construction work. | High mechanical aptitude. Sheet metal, foundry, pipefitting, carpentry, mathematics, geometry and chemistry valuable. | Fire fighter, welder, plumber, shipfitter, blacksmith, metallurgical technician, carpenter. |
| **Electrician's Mate** | Maintains power and lighting equipment generators, motors, power distribution systems, other | Aptitude for electrical and mechanical work. Electrical, practical and shop mathematics, | Electrician, electric motor and electrical equipment repairer, armature winder, radio/TV |

| | electrical equipment; rebuilds electrical equipment. | and physics. | repairer. |
|---|---|---|---|
| **Electronics Technician** | Maintains all electronic equipment used for communications, detection ranging, recognition and countermeasures, world-wide navigation systems, computers and sonar. | Aptitude for detailed mechanical work. Radio, electricity, physics, algebra, trigonometry and shop valuable. | Electronics technician, radar and radio repairer, instrument and electronics mechanic, telephone repairman. |
| **Fire Control Technician** | Operates, tests, maintains and repairs weapons control systems and telemetering control systems and telemetering equipment used to compute accuracy of naval guns and missiles. | Perform fine, detailed work. Extensive training in mathematics, electronics, electricity and mechanics. | Radar or electronics technician, test range tracker, instrument repairer, electrician. |
| **Gunner's Mate** | Operates and performs maintenance on guided-missile launching systems, rocket launchers, guns, gun mounts; inspects/repairs electrical, electronic, pneumatic, mechanical and hydraulic systems. | Prolonged attention and mental alertness, ability to perform detailed work. High aptitude for electrical and mechanical work. Arithmetic, shop math, electricity, electronics, physics, machine shop, welding, mechanical drawing. | Gunsmith, locksmith, machinist, instrument repairer, hydraulics, pneumatic or mechanical technician, small appliance or test equipment repairer. |
| **Health Services** | Administers medicines, applies first aid, assists in operating room, nurses sick and injured, and assists dental officers. | Hygiene, biology, first aid, physiology, chemistry, typing and public speaking. | Practical nurse, medical, dental, or X-ray lab technician, pharmacist, emergency medical technician. |
| **Machinery Technician** | Operates, maintains and repairs ship's propulsion, auxiliary equipment and outside equipment such as steering, engine, | Aptitude for mechanical work. Practical or shop mathematics, machine shop, electricity and physics valuable. | Boiler house repairer, engine maintenance, machinist, marine engineer, turbine operator, engine repairer, air conditioning and |

| Career Fields | Duties & Responsibilities | Qualifications | Examples of Civilian Jobs |
|---|---|---|---|
| | refrigeration/air conditioning, steam equipment. | | refrigeration repairer. |
| Marine Science Technician | Makes visual/instrumental weather and oceanographic observation; conducts chemical analysis; enters data on appropriate logs, charts, and forms; analyzes/interprets weather and sea conditions. | Ability to use numbers in practical problems. Algebra, geometry, trigonometry, physics, physiography, chemistry, typing, meteorology, astronomy and oceanography useful. | Oceanographic technician, weather observer, meteorologist, chart maker, statistical clerk and inspector of weather and oceanographic instruments. |
| Public Affairs Specialist | Reports, edits, copyreads news; publishes information about service people and activities through newspapers, magazines, radio and television, shoots and develops film and photographs. | High degree of clerical aptitude. English, journalism, typing and writing experience helpful. | News editor, copyreader, script writer, reporter, free lance writer, rewrite or art layout person, producer, public relations, advertising, photographer. |
| Quartermaster | Performs navigation of ships, steering, lookout supervision, ship-control, bridge watch duties, visual communication and maintenance of navigational aids. | Good vision and hearing and ability to express oneself clearly in writing and speaking. Public speaking, grammar, geometry and physics helpful. | Barge, motorboat, yacht captain, quartermaster, harbor pilot aboard merchant ships, navigator, chart maker. |
| Radarman | Operates surveillance and search radar, electronic recognition and identification equipment, controlled approach devices and electronic aids to navigation; serves as plotter and status board keeper. | Prolonged attention and mental alertness. Physics, mathematics and ship courses in radio and electricity helpful. Experience in radio repair is valuable. | Radio operator (aircraft, ship, government service, radio broadcasting), radar equipment supervisor, and control tower operator, air traffic controller. |
| Radioman | Operates communication, transmission, reception, and | Good hearing and manual dexterity. Mathematics, physics | Telegrapher, radio dispatcher, radio/telephone operator, |

| | | | |
|---|---|---|---|
| | terminal equipment; transmits, receives and processes all forms of military record and voice communications. | and electricity desirable. Experience as amateur radio operator helpful. | computer operator. |
| **Sonar Technician** | Operates electronic underwater detection and attack apparatus, obtains and interprets information for tactical purposes, maintains and repairs electronic underwater sound detection equipment. | Normal hearing and clear speaking voice. Algebra, geometry, physics, electricity and shopwork desirable. Experience as amateur radio operator. | Oil well sounding device operator, radio operator, inspector of electronic assemblies, electronic technician, electrical repairer. |
| **Storekeeper** | Orders, receives, stores, inventories and issues clothing, foodstuffs, mechanical equipment and other items, payroll clerk. | Typing, bookkeeping, accounting, commercial math, general business studies and English helpful. | Sales or shipping clerk, warehouse worker, buyer, invoice control clerk, purchasing agent, accountant. |
| **Subsistence Specialist** | Cooks and bakes, prepares menus, keeps cost accounts, assists in ordering provisions, inspects foodstuffs. | Experience or courses in food preparation, dietetics, and record keeping helpful. | Cook, pastry chef, steward, butcher, chef, restaurant manager. |
| **Telephone Technician** | Installs, operates, maintains and repairs all telephone, telegraph and teletype equipment, switchboards, public address systems and inter-office communications systems. | Aptitude for electrical and mechanical work, use of numbers in practical problems. Previous electrical experience helpful. | Electrician, electrical equipment inspector and many jobs which are in the civilian field of telephonic communications. |
| **Yeoman** | Clerical and secretarial, typing, filing, operating office and duplicating equipment, preparing and routing correspondence and reports, maintains personnel records and offical publications. | Same qualifications required of secretaries and typists in private industry; English, business subjects, stenography and typewriting helpful. | Office manager, secretary, general office clerk, administrative assistant, legal clerk, personnel manager, court reporter. |

# Index